The History of Western Glow Guernseys

A History with Guernsey Cattle

by Kirt Sloan

1st Edition

Table of Contents

The Story of Western Glow 1
   The Farm 6

Chapter One
The Foundation Sire 12
Hilltop Butterfat Count 12

Chapter Two
The Breeding Program 15
The Langwater Influence Brought Forward 15
   McDonald Farms Developer 274206 b: November 6, 1938..................16
   Silver Forest May Royal 327032   b: July 2, 1942..................................17
   Fritzlyn Jeanette's Flash 475548   b: December 28, 1950......................18

Chapter Three
The Whole Maryann 20

Chapter Four
The Influence of Virtue 30
   Western Glow Fames Virtue 30
   One Cow Three Sons 30
   The Sons..32
   Other sires from the V Fame bloodline...34
   Clovelly Top Hornet..34
   Virtue of Aldarra...35

Chapter Five
The Duchess Family 36
   A Different Branch of the Family Tree..36
   Western Glow Travelers Duchess..36
   Western Glow Fond Gay Duchess Ex-94..37
   The Daughters of Fond Gay Duchess..41

    Western Glow M.B. Duchess VG 87 @ 9-02 41

    Other daughters of Fond Gay Duchess 43

    Other Daughters of Travelers Duchess 45

Chapter Six
The Story of Darimost 46
6,666 daughters in 1,120 herds 46

Chapter Seven
Sun Glow Bev 54
by Lake Louise Patsy's Prince 54

Chapter Eight
And there were others 59
J.L Princess Ex92 59
    The other Lake Louise Patsy's Prince daughter 59

    The Third Patsy's Prince daughter 61

Western Glow Brilliant Mae 62
    Beautifully uddered and balanced 62

    Western Glow Champions Mardey VG86 63

Chapter Nine
The Golden Years 64
    The Washington State PDCA Hall Of Fame Award 64

Chapter Ten
The Future in Black and White 66
    The changing of the guard 66

    The end of an era 67

    Western Glow Farms Bow, Washington circa 1936 67

    The Western Glow prefix (Holsteins) 70

Chapter Eleven
Western Glow - Part 2 71

We bought a Guernsey 71
Going Old School 75
Dedication 78

The Farm......

This is an oral/written history of events, some that happened long before I was born. I have done my best to document the facts and stay true to them, but as all good stories that I heard growing up, there may be some things that were exaggerated. So this is kind of a historical novel....enjoy.

I have been very fortunate in my life. One of my biggest breaks was to have Ben Friedrichs as my grandfather and to have grown up with, and been infused with, his love for good cattle and how to take care of them with great skill and devotion. He was part of a team with his brother Jake Friedrichs that bred and developed the herd of Guernseys know to the world by the Western Glow prefix. Jake was one of the best caretakers a cow could have. His calm demeanor and his ability to feed a cow to her potential using local forages was unequalled. He loved the cows, his garden, and hunting. He was integral to the success of the farm and maximizing the results of the breeding decisions. Ben was the breeder and the promoter, he was the face that other breeders saw at shows and sales, he loved people and he loved to help young people with project calves and judging. The sum of the team is what made the difference in achieving the performance results. This herd saw World Records for production, top prices in sales over the United States and Canada, and many All American awards for animals bred by Western Glow and out of the bloodlines that were developed there.

This journey they were on started in Germany in a farming community along the North Sea. They grew up farmers as did the generations before them. As young children they experienced the first World War and the hardships that went along with that. He told stories of them roasting barley to make coffee. In 1924 during the peak of the hyper inflation in Germany, Ben at 20 years old, got on a ship and sailed to the United States leaving everything behind. His first job was milking 28 cows three times a day in a test barn in Burlington, Washington. He saved his money and four years later in 1928, he returned to Germany to bring back his brother Jake who was 18 and his sister Emmie. Emmie stayed and settled in Iowa on their journey across the United States to get back home to Burlington. Grandpa brought a female purebred German Shepard back from Germany on that trip and the story goes she saved them from a dangerous encounter with some thieves in the midwest along the way. When they got home he showed the dog and sold her puppies to buy their first cow.

In 1928 the brothers rented the farm they would later purchase in 1932. It was an 80 acre farm in Bow, Washington. The land was flat as it had been tide flats before the dikes were put in and the land drained in the early 1900's. The soil was good though drain tile made it better to access the pastures earlier in the spring. Grass grew tall and cows did well in the coastal zone with the moderate climate. There are few places on earth with this climate, near perfect for dairy cattle and milk production.

The farm was named Western Glow, as at sunset the sun reflects off the water behind the dikes and it creates a glowing band on the western horizon at high tide. The story went as they were struggling to come up with a prefix at the kitchen table one of those brilliant sunsets was in process, someone said "That's it! Western Glow." From the farm is a magnificent view of the mountains and hills in the back ground and the San Juan Islands to the West. This place is one of the most beautiful places on earth to catch a sunset; with the islands and the Chuckanut Bay in the foreground. Clouds from the frequent rains turn the skies into exquisite beauty. Getting the cows in from the pastures in the morning as the sun rose and the fog lifted was almost a spiritual experience that is hard to put into words. They paid $10,000.00 for the farm, a small price for the pallet that they would paint the future upon.

There were years of struggle and years of good times followed by severe financial trials. They ventured into other business leveraged on the success of the herd, and that did not always go well. In talking with Jake in his later years he told me that before the adoption of artificial insemination, they sold every bull calf that hit the ground for $750.00 or more to become herd sires of other farms. Considerable money for the 1940's. They hauled milk to Bremerton, Washington during the War. A typical day started at 3:00am and ended late in the evening. They worked hard but did well. They owned interest in a fish cannery in LaConner, Washington that sold canned fish to the military during the War. They thought the War would last many years. The War ended abruptly and they lost a significant amount of money. The cows had to carry the load. They also with other Guernsey producers had a creamery in Everett, Washington called Arown Milk Farms that sold fluid milk and ice cream. The Class 1 Base program came into being and they no longer had the higher Class 1 prices for their milk. They had to pay into the pool where the income from Class 1 sales were blended down with the manufacturing Classes of milk, the profits were gone. Grandpa paid some of the smaller producers out of his pocket when the business failed. In total these two set backs cost in the realm of $250,000.00 a huge sum for the times. Years later I met with the family attorney and he said that he had recommended that they file bankruptcy. Grandpa refused and over the next 30 years he paid all the debts to settle the accounts.

Grandpa had the knack of finding the diamond in the rough and as a child I accompanied him on some of these trips to small Guernsey farms in the Pacific Northwest. There were two types of trips. One was to purchase cattle, usually a small herd where we would keep the top couple of cows and drop the rest off at the stockyards on the way home. And the other was to drop off a young bull and pick up the older bull from the last visit as dairy farmers got rid of the older bulls. Of course in most cases the older bull was dropped off at the stockyards on the way home. They had bulls out all over the area as they were sought after herd sires with Western Glow bloodlines.

Another thing Grandpa loved to do was show cattle. For over 30 years each summer he did the Prairie circuit in Alberta, Canada. In the early years cattle were loaded on baggage cars in Mission, British Columbia and taken through the Rockies to Calgary, Edmonton, Red Deer, and Lethbridge during the 4 week schedule in July each year. There were 20 head on a 90 foot baggage car, usually 2 herds on each car with 10 head on each end of the car. The tack and feed were stored in the center. The water was stored in 55 gallon lard drums and water was dipped out to water the cows along the trip. We slept on the hay bales. One of the first things we learned was to not to throw manure or milk out of the train going through towns. Before the lesson I saw my brother throw a pile of manure out the door and it landed on the hood of a car at a crossing. The other thing you learned was to tie the bulls against the ends of the cars…and to tie them short so they could not spin the wheel on the emergency brake for the train. In the baggage car the wheel for the brake was inside the car. One of the bulls tied too loose did that and that really pissed off the train crew. The tunnels in the Rockies were long and winding as we climbed the mountain passes. We had a cow calve on the way home as we went into the tunnel, never have i seen it that dark and the diesel fumes got a little thick towards the end of it. We were stuck sitting by the cow and calf as we could not see to get back to the middle of the car. As an 11 year old I made the trip one summer showing a heifer Grandpa gave me using the prize money to buy school clothes. This was quite a trip with the beautiful scenery of the Frazier River canyon, the Rocky Mountain lakes, and the open prairies. The train would roll first into Calgary where like the circus we unloaded the cattle from the train car and walked them to the barns to get ready for the show that week. We watched the Chuckwagon Races at the Calgary Stampede through a hole in the fence and slept in a dormitory with hundreds of snoring old farmers. The next week we would go to Edmonton and repeat the whole process. It was there in 1969 that we watched the Apollo 11 moon landing on a small black and white TV that was sitting on someone's tack box in the cow barn. Usually we would sell one of the young bulls along the way to Canadian breeders. At Lethbridge that year the Canadian leader Pierre Trudeau walked through the barns and we got to see him. During those years many great Canadian herds of all breeds showed the Prairie circuit as the prize money was around $125.00 for a first place ribbon and a 10 head string could knock down $2,000.00 on a good show day with group classes included. We usually did well and that $7,000 to $8,000 in prize money was motivation to run the circuit and make good money.

In the early days the expositions paid of the feed and straw of the cattle shown as well as the rail freight. One of Grandpa's favorite classes to win was Best Three Females, Bred and Owned.

Selling cattle to wealthy breeders Back East who wanted show cattle also brought in good revenue. Atherton Hobler of Woodacres in Princeton, New Jersey bought several All Americans from Western Glow starting in the 1960's with Western Glow Fond Babbette the All American 2 year old in 1962. There were others too. Henry Venier of High Meadows in Brentwood, New York bought Candy Jewel a Darimost daughter Grandpa purchased and resold. She was All American Aged Cow in 1968.

W.E. Boeing also came to the farm one morning in the 1940's. Grandpa told the story many times "A fancy car drove up to the barn as I was in the barn brushing the cows after milking, and a well dressed gentleman walked up and asked if I was Ben Friedrichs. I said I was and he introduced himself as Bill Boeing and said 'Some people I have talked to said I should come to see you about buying some good cows'." They walked through the cows and after negotiating a deal Boeing bought all the young stock on the farm from baby calves to springing heifers for $1,200.00 a head. It was only 25 head on a small herd but a spectacular deal for the times. Later in the week Grandpa hauled the first load to Aldarra Farm near Seattle and went back home to get the second load. Boeing met him as he was dropping the end gate of the truck and said "Hey I have been talking to some other people and they say you might be trying to screw me on this deal." Grandpa said "Mr. Boeing, that is not a problem if you want to call off the deal I will take these home and come and get the rest, no hard feelings." Boeing replied "Well if you feel that way about it it must not be too bad a deal." And Boeing shook his hand saying "From now on I am Bill and you are Ben." They owned cattle and bulls together and when Boeing passed away Grandpa helped market cattle to settle the estate. He and my Grandmother also were guests on Boeings yacht.

There were always stories and riding with Grandpa to shows and sales with his friends in later years I learned the art of telling adult jokes that I was not supposed to hear. And on those show days most tack boxes had a bottle of whiskey in them just in case your cow won Grand Champion on show day. The industry has changed but the breeders still make the good ones and the guys with money still buy them.

One of my deepest regrets was that i was not aware enough or old enough to ask my Grandpa Ben about the history and the events that not only brought him to the U.S. but also how he went about selecting the foundation animals and the selection of the matings for the cow families that ensued. I asked him one time "What was the secret to your success?" He replied "Well, we just kind of got in a rut and stayed there."

Perseverance through the Great Depression, World War II, financial adversity, a changing industry and society charted the course for his life. He as born in 1904, one year after the airplane was invented and passed away in 1982 in the early years of the computer age. He started milking cows by hand in 1924, farmed with horses through the 1930's. In 1951 they put up the first two Harvestore silos (the blue ones) in the Western U.S. In 1974 they put in Universal automatic milking machines, yet through all of that they stayed in their rut and created a legacy with Western Glow Farms.

In 982 my Grandpa passed away and in his desk were pictures taken over the years by Harry Strohmeyer, Danny Weaver, and others. Some had his hand writing on the back of them and others had my writing from childhood on the back of a few of them. I would go through the pictures in the drawer as he sat at his roll top desk doing book work and record keeping of the registration papers and breeding records. I would ask him who the cows were and who they were out of. He would tell me and I would write on them. Also in the papers were 4 six generation pedigrees from the 1930's to the 1950's. I looked at them every few years but never realized the importance of them nor how they all fit together until a bout a year ago. There was a reason he kept those pictures and those pedigrees over the years, they were important to him. Then all the pieces started to fit....

The passion was reignited this past year when a friend on Facebook Cliff Shearer told me about the CDN (Canadian Dairy Network) website cdn.ca . It has an amazing database full of historical data. I did a search for Western Glow under the Guernsey breed and found over 100 bulls with the prefix as well as over 100 females going back to the 1930's. I started searching pedigrees and kept hitting a dead end on a bull with the registration number GUUSAM218620. Walking down the hallway one morning I looked at the framed 6 generation pedigree for Hilltop Butterfat Count that my mother gave me about 35 years ago.....and I found my answer.

A couple of weeks later I was on Facebook and saw a picture of a huge old white barn with the caption Hilltop Farm. I read the information on it and saw that is was at one time owned by George Hendee, another trip down the hall and sure enough his name was on the pedigree as the breeder of Hilltop Butterfat Count. More on that story later.

Another event that stirred my enthusiasm was a picture posted on Facebook of a Guernsey cow being loaded on an airplane many years ago. I had heard stories about a cow Grandpa sold in Chicago in the 40's but I did not put the pieces together. An old friend of mine Seth Spenser from Oregon commented "I thought you would remember that was Western Glow Butterfat Maryann." and he told me the story about the picture. So I got on CDN and the rest you could say is history....

Another gem that I can attribute to Cliff Shearer was the mention in a Facebook post about a book he had on Langwater Guernseys, I had seen the name in the old 6 generation pedigrees so I ordered the book online from abebooks.com they do historical reprints. It was another piece to the puzzle.

In 1974 at the Western National Guernsey Show

Best Three Females - Bred and Owned

l to r: Dorthy Crask, Corney Boon, Ben Friedrich Sr., Ben Friedrichs Jr., Kirt Sloan, Fred Rehm, Teresa Bosch

Chapter One
The Foundation Sire
Hilltop Butterfat Count

In 1935 Grandpa made a trip East to Hilltop Farm in Suffield, Connecticut to purchase a herd sire for the herd they had assembled. His choice was Hilltop Butterfat Count a line bred son of Hilltop Butterfat who had two crosses to Imp. Primrose's Butterfat. Count's Dam was Hilltop Butterfat Maid an own daughter of Imp. Primrose's Butterfat that produced over 830 pounds of fat and was twice a Class Leader for production.

Hilltop Butterfat Count

He had purchased the key bull, an inbred outcross. Below is the six generation pedigree of Count the printing faded over the past 80 years but it is still just barely readable.

Throughout history many great herds have been created with this proven concept of bringing an inbred sire from an outcross bloodline into the existing herd. This does two things, it concentrates desirable traits that you are selecting for to get them to transmit a consistent breeding pattern, and two if you have a line bred herd of a differing bloodline you can create an new line to work off of that has a higher likelihood of transmitting consistent traits in their offspring.

The rest of that story on that transaction, Grandpa Ben shared and loved telling. After he bought the bull for $1500.00 a lot of money during the Depression he said "I had to hurry home after I wrote the check for the bull, because it was not good when I wrote it. I went to the bank to see the banker and he was not going to loan me the money. I told him if that was the case I needed to find another bank. The guy looked at me and took the picture of the bull I had with me and said 'Alright if you feel that strongly about it I will loan you the money' and he took the picture and put it under the glass top on his desk. It was there until he retired from the bank."

Hilltop Butterfat Count had 9 milking daughters the first time they classified for type at Western Glow. 7 of them went Excellent. He knocked it out of the park. His daughters would go on to set World Records for Production in the Guernsey breed as we as transmit exceptional type for many generations. And they were exceptional brood cows that transmitted it on through their sons and daughters. Most of what is written in this book had it's foundation in offspring of Hilltop Butterfat Count……but there is more to where Count came from.

Hilltop Farm was owned by a man named George Hendee. He had spend his early years racing highwheel bicycles and at 16 in 1882 he was the U.S. Amateur Champion and held the title until 1886. During his career he won 302 out of 309 races he entered. In his last race in 1897 he won the World Championship. In 1900 he met Carl Hedstrom as he admired the performance of Hedstrom's pacer and in 1901 he recruited him to join his new company as chief designer and engineer to build a "motorized bicycle for the masses". George Hendee was the co founder of Indian Motorcycles. Hendee Manufacturing became the largest employer in Springfield with 3200 employees. Hendee Manufacturing Co. by 1913 was the largest motorcycle company in the world. Hendee stepped down as President in 1915 and as Chairman in 1916 to retire to Hilltop Farm. Hendee Manufacturing Co. became Indian Motorcycle Co. in 1923. [1]

Hilltop Butterfat Count may have come from a famous herd with a wealthy owner but he made a name for himself on a small farm in Bow, Washington through his daughters and their offspring.

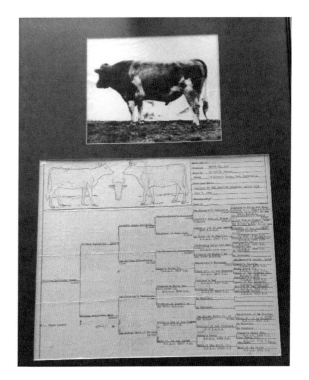

His Daughters: Partial List

Wesclair Pansy's Flower

WayLayne Butterfat Countess

Western Glow Butterfat Lou

Western Glow Butterfat Duchess

Western Glow Butterfat Lark

Western Glow Butterfat Miss

Western Glow Butterfat Maryann

---

[1] From http://hilltopfarmsuffield.org/about/the-farm/george-mallory-hendee

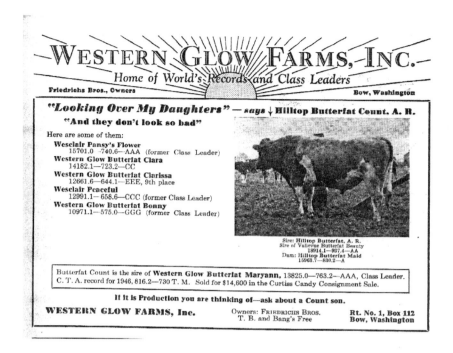

One of the ads in the Guernsey Breeders Journal

Hilltop Farm in Suffield, CT

Chapter Two
The Breeding Program
The Langwater Influence Brought Forward

One of the greatest Guernsey herds to ever walk the planet was Langwater. [2] From the early 1900's to his untimely death in 1921 Fredrick Lathrop Ames did a masters work of assembling and breeding a herd that was truly extraordinary. He was a Harvard graduate from a wealthy family. His Grandfather made his fortune selling picks and shovels to the gold miners during the Gold Rush in California. His company was Ames Manufacturing. Mr Ames could have chosen any path but his dedication the the Guernsey breed will be forever appreciated.

Pictured is Pansy's Maryann at 11 1/2 years sired by Langwater Traveler a son of the incredible Langwater Holliston a son of an own daughter of Imp. King of the May 9001. (The kingpin of the Langwater program) She was one of the original females at Western Glow and her mating to Hilltop Butterfat Count would change things in a big way in the not to distant future. Other than Count most of the herd sires added over the years were heavily influenced by Langwater breeding. The Langwater Guernseys had beautiful udders with great frame and substance. The bulls were strong and masculine and the females refined yet had great substance. That is something lost on todays breeders, too many males are refined and weak which leads to lack of substance in their offspring. Frailty is often mistaken for dairy character. I hope you get a chance to read the book on Langwater listed in the footnotes it is an incredible documentation of that great herd and breeding program.

With the Count daughters established as the maternal foundation at Western Glow the search was on for the future herd sires to compliment them. Grandpa Ben always went back to bulls rich in Langwater blood starting early on and until the late 1950's. We will go through some of those key sires now.

---

[2] From the book Langwater Guernseys by William H Caldwell and C.W. Barron

McDonald Farms Developer 274206 b: November 6, 1938

A Foremost Prediction son out of Rockingham May Princess (pictured) who had Langwater Holliston on both her dam and sires sides. there are 25 crosses to Langwater animals in a 6 generation pedigree of Developer. He was used heavily at Western Glow. J.M. McDonald of McDonald Farms was a manager for J.C. Penny and had his own herd. also. J.C. Penny of course was the owner of Foremost Guernseys. Both herds were influential.

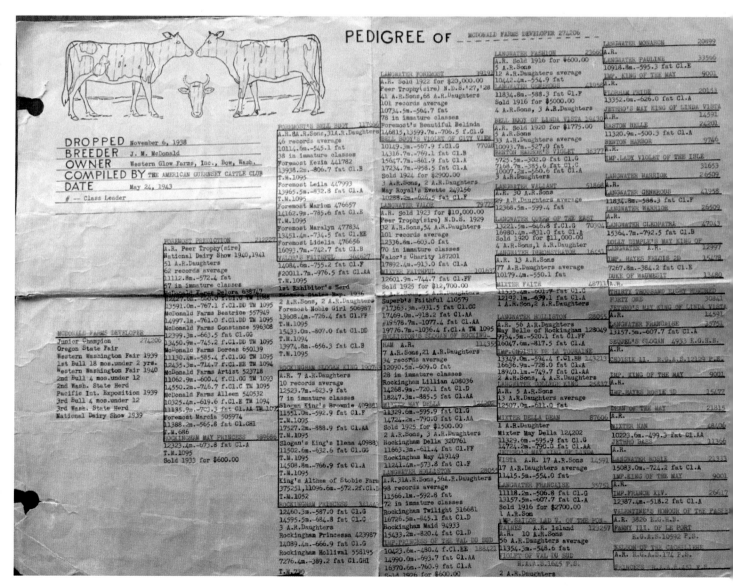

Silver Forest May Royal 327032   b: July 2, 1942

Though his own daughters were decent cows. He was owned in partnership with W.E. Boeing of Boeing Airplane fame and also used at his Aldarra Farms. The true influence was through his sons. He was an intensely Langwater bred son of Langwater May Royal and out of his dam Langwater May Blossom who made 724 pounds of fat and sold for $3,500.00 in 1940. May Blossom's dam and sire were both sired by Langwater Pharaoh. There are 41 shots of the Langwater prefix in this special sire. He worked very well on the Developer daughters as well as the Counts.

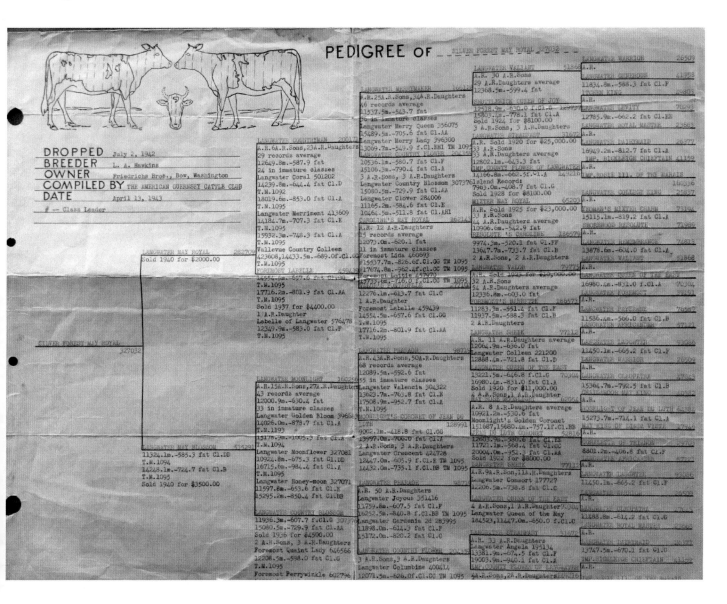

His son Western Glow Royal Fame was the sire of Western Glow Fames Virtue the dam of three key bulls that sire many All Americans in milking form. Gayoso View Top Command, Henslee Farm V Fame and Henslee Farm Viscount.

Fritzlyn Jeanette's Flash 475548   b: December 28, 1950

Flash may have been late to the party buy did he ever make an entrance! He was one of the last of the foundation sires with a considerable amount of Langwater blood in the pedigree. He had a huge influence on the herd through his offspring and the next generations.

He had 17 shots of Langwater prefix in the 6 generation pedigree and his dam and paternal grand dam were both out of a son of Langwater Vagabond a bull who sold for $16,000.00 in 1943. His son Antietam Gypsy King, the maternal grandsire of Flash as well as the sire of his paternal grand dam, had daughters with records over 1000 pounds of fat.

He sired several All Americans in milking form and transmitted smoothness and high test.

His offspring worked well in line breeding in the generations that followed.

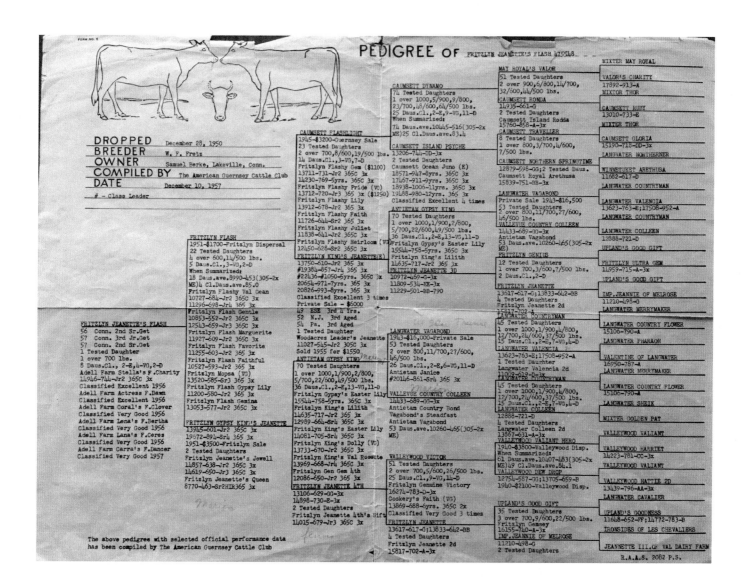

Flash would have breed wide influence with the most widely used bull with the Western Glow prefix.....

Western Glow Darimost a bull that saw heavy service at All West Breeders and sired many All Americans in milking form.

l: Flash's best daughter was Western Glow Fond Gay Duchess EX 94 2x All American

She made over 1000 pounds of fat with a lifetime 5.4% butterfat.

r: Western Glow Fond Babette EX 90 was All American 2yr old in 1962. Another Flash daughter that was owned by Woodacres in Princeton, New Jersey.

Chapter Three
The Whole Maryann

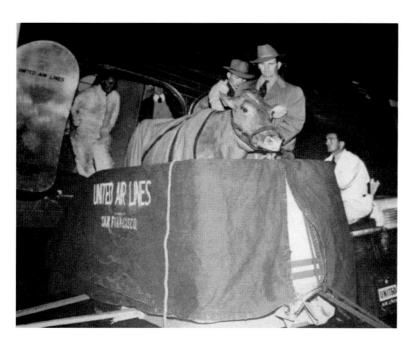

In 1946 a year after the war had ended things were getting back to normal and people had big ideas after 15 years of difficult times. Grandpa consigned **Western Glow Butterfat Maryann** to the 1946 Curtis Candy Sale that fall in Chicago. She had made a Class Leader Record AAA 13,825 milk 5.5% 763 fat She was an own daughter of **Hilltop Butterfat Count** out of **Pansy's Maryann**. Foundation dam. Foundation sire. She was 8 1/2 years old.

The story goes…Grandpa had consigned her to the sale and had second thoughts. He went to the sale in Chicago ran by Louis Merryman. Grandpa arrived early because he had something to tell him. He had left Maryann at home, he thought a lot of her and wanted to keep her. Louis Merryman had other ideas. The conversation went like this. Louis said "I don't care what you have to do but get her here, we have people that want her!" Probably more colorful language was involved but you get the point. Grandpa knew Bill Boeing by this point and made arrangements to have Maryann flown to Chicago on United Airlines. She was hauled to Seattle loaded on an airplane and flown through San Francisco to Denver where she was milked and then flown onto Chicago for the sale. I had heard the story of the cow going to the sale in Chicago but never saw the pictures until David Cochard sent me the scanned pictures earlier this year out of a 1946 Guernsey Journal. They are priceless. The entire sales report is here. Maryann left a son **Western Glow Maryann's Royal** by **Silver Forest May Royal** that was used at Western Glow and other herds. Jimmy Olsen had a Maryann's Royal daughter that would become the dam of **Western Glow Darimost** a son of **Fritzlyn Jeanette's Flash.** The foundation was solid and the market demand was very good for Western Glow cattle.

Western Glow Butterfat Maryann, top of the Curtiss Candy Sale. She was consigned by Friedrichs Brothers, Bow, Wash., and purchased by M. C. Fleming, Troutdale, Ore., for $14,600. In the background, M. C. Fleming and Mr. & Mrs. Ben Friedrichs.

Quality and Demand Set High Prices
Curtiss Candy Sale Breaks All Records

THREE new auction sale records were set at the Curtiss Candy Guernsey Sale held at Curtiss Candy Company Farm, Cary, Ill., on October 7. 1.—An all-time auction sale average was established when 50 head sold for $144,400, an average of $2,888. 2.—An all-high price was paid for a bred heifer—$13,500 for McDonald Farms King's Kella. 3.—The highest price was paid for an open heifer, Curtiss Candy Noble Deborah, that sold for $10,500. With the rapid increase and advancement of Guernseys in America, it took 26 years for an average of an auction sale to surpass that set by F. Lothrop Ames on September 21, 1920, when 51 head sold at Langwater Farm for $146,125, an average of $2,865.

This has been a year of record prices and it was very fitting that the Curtiss Sale should be the one to top all averages for the breed. Animals were selected from 40 breeders in 19 states, with an eye of assembling the best possible offering. No place could have been more fitting than Curtiss to hold this record-breaking sale. The layout of the buildings, accommodations for the consignments, the sales pavilion and the hospitality of Mr. & Mrs. Otto Schnering, owners, and Delbert Kingston, superintendent, set the stage for the great event.

Sunday afternoon a parade and show of the animals was held, followed by all being the guests of Mr. & Mrs. Schnering for a buffet supper. No efforts were spared by the Curtiss organization to see that everything was in readiness for Monday.

As the time for the sale approached, the 1,200 to 1,500 breeders and spectators gathered early and waited, with ever-increasing anticipation, the outcome of the sale. One seemed to sense in the atmosphere around the barn and the ring that records were to be broken. Old-timers who have attended many sales said they were getting a real thrill.

One of the big gambles in the sale was the bringing of six head from

Guernsey Auctions that Have Averaged $2,000 or More

Sale	Head	Date	Average
Curtiss Candy Guernsey Sale	50	1946	$2888.00
Langwater Sale	51	1920	2865.00
Cherub-Levity Sale	48	1924	2798.96
Langwater Dispersal	96	1922	2738.85
The Guernsey Sale	66	1944	2398.48
The Guernsey Sale	45	1943	2393.33
Argilla Farm Dispersal	91	1946	2335.16
Second Combination Sale	83	1919	2172.00
The Guernsey Sale	52	1946	2118.75

Ring scene at opening of sale as Western Glow Butterfat Maryann sold for $14,600.

Washington and Oregon. This is the first time that such a large consignment has been sent East.

To stimulate added interest, Western Glow Butterfat Maryann was flown from her home in Washington, via San Francisco, to Chicago. She was also given first place in the sales order and after the regular introductions to the sale, she was led into the ring. As this great bodied, eight-year-old cow, paraded around the ring a hush fell over the crowd. Louis Merryman spoke of her pedigree and the work of the Friedrichs Brothers in their farm home, Bow, Wash. Bidding opened at $3,000 and continued to $12,000 at $1,000 increases, then settled to $500 and the last bid was $100, making the selling price $14,600 paid by M. C. Fleming, Troutdale, Ore. The cow had come East, been appraised and was returned to the West Coast. A farmer-breeder had bred the top of the sale and a farmer-breeder had purchased the top of the sale. The contending bidders were Lloyd Wescott and R. A. Loeb, both of Clinton, N. J., who hoped to own her jointly. Several other breeders were in for a while.

Western Glow Butterfat Maryann was by Hilltop Butterfat Count, a grandson of old Imp. Hilltop Butterfat's Clara that held a world's record in AAA, and also a grandson of Imp. Primrose's Butterfat. On the bottom side, "Maryann" was out of Pansy's Maryann that has a record in A of 14781.4 lbs. of milk and 768.4 lbs. of fat. "Pansy" was sired by Langwater Traveler and out of Waldo Hills Pansy. There was great production on all sides of this really good cow.

McDonald Farms King's Kella, dropped on January 10, 1945, was the second high bringing $13,500 from Charles B. Bolton, Franchester Farm, Lyndhurst, Ohio. The contending bidder was the Curtiss Candy Company. "Kella" was sired by Myhaven King, a son of Langwater King of the Meads and out of Green Meads Thelma. The dam of "Kella" was McDonald Farms Della, a daughter of Foremost Prediction and Douglaston Lady Augusta. "Kella" came from a great cow family and it is fitting that she should bring the all-high price paid for a bred heifer.

One of the high spots in the sale was the appearance in the ring of Quail Roost Noble Primrose, the recent grand champion at the Dairy Cattle Congress, and her daughter, Curtiss Candy Noble Deborah, that was being consigned to the sale. "Deborah" was dropped on April 2, 1946, and was purchased by Mr. & Mrs. F. L. Weyenberg, Wey Acres, Thiensville, Wis., for $10,500. The contending bidders

McDonald Farms King's Kella, that was consigned by J. M. McDonald, Cortland, N. Y., and purchased by Charles Bolton, Lyndhurst, Ohio, for $13,500. "Kella" brought the all-high price for a bred heifer.

Curtiss Candy Noble Deborah that was purchased by Mr. & Mrs. F. L. Weyenberg, Thiensville, Wis., for $10,500, the highest price ever paid for an open heifer. In the background, Mr. & Mrs. F. L. Weyenberg, owners of Wey Acres, Thiensville, Wis., purchasers; Delbert Kingston, superintendent of Curtiss Candy Company Farms; Mr. & Mrs. Otto Schnering, owners of Curtiss Candy Company Farms, consignors; Otto C. Kline, manager of Wey Acres, and Merle Campbell of Curtiss Candy Company Farms.

Afton's Golden Marie, consigned by W. H. Scott & Gordon W. Bridges, Hunterdale Farm, Franklin, Va., and purchased by Curtiss Candy Company for $10,000. In the background, Delbert Kingston, Mr. & Mrs. Schnering and Merle Campbell.

were Quail Roost Farms, Rougemont, N. C., and J. C. Penney, Emmadine Farm, Hopewell Junction, N. Y. Mr. Penney placed the last bid before the sale. This was another record in price, for no unbred heifer had ever been sold for that much money. She was by Curtiss Candy Levity Chum, that was sired by Green Meads Levity King and out of Douglaston Princess Charlotte. The dam, as stated, was Quail Roost Noble Primrose that was by Cesor Noble Maxim and out of Quail Roost Royal Primrose.

Afton's Golden Marie, consigned by W. H. Scott and Gordon W. Bridges, Hunterdale Farm, Franklin, Va., was the next high animal in the ring. Curtiss Candy Company paid $10,000 for this five-year-old and the contending bidder was George Watts Hill, Quail Roost Farms. "Marie" was sired by Bournedale Golden Steadfast that, in turn, was by Bournedale Golden Majesty out of Bournedale Memoir. On the dam's side she was out of Afton's Millie, a daughter of Twenty Grand and Springwater's Millie.

McDonald Farms Dairymaid, another daughter of Myhaven King out of McDonald Farms Alleen, a daughter of Foremost Prediction, was purchased by Curtiss Candy Company for $8,000. The contending bidder on this cow was Franchester Farms, Lyndhurst, Ohio, that was represented by C. R. Huston.

Two other head brought $5,000 or better. Riegeldale Babette, consigned by Riegeldale Farm, Trion, Ga., sold for $5,400 to Charles G. Lang, Glenarm, Md. Quail Roost and Franchester Farms were the contending bidders. Also, $5,000 was paid for Curtiss Candy Levity Hero, a son of Green Meads Levity King out of Sun-Blest Farm Lotus. "Hero" was two years of age and purchased by William Y. Gilmore, Lookout Valley Farm, Lake Geneva, Wis. Hooper & Strever, West McHenry, Ill., were the contending bidders.

Another Western cow, Dorien's Lily, a four-year-old, consigned by Meyer Brothers, Olympia, Wash., was purchased by W. B. Vilter, Hartland, Wis., for $4,700. She was sired by Boulder Bridge Philosopher out of Bonny Brook Dorien. The contending bidders were Boulder Bridge Farm at $4,600 and Quail Roost Farms at $4,000.

Kent B. Hayes, Meadow Lodge Farms, Oklahoma City, Okla., consigned Meadow Lodge Gay Flute, a daughter of Meadow Lodge King's General out of Argilla Flute. "Gay Lute" was purchased by Weston Howland, Brattleboro, Vt., for $4,600. The contending bidders were Don C. Roberts, Mahomet, Ill., and John S. Ames, Langwater Farm, North Easton, Mass.

Eight bulls were included in the consignments that sold for $16,100, or an average of $2,012. Only three of the animals consigned brought less than $1,000, with 42 head of females averaging $3,054.

The highest buyer was Curtiss Candy Company that purchased 10 head for $31,100. Shoal Falls Farm, Inc., Hendersonville, N. C., secured four head for $5,200, while two other buyers purchased three head each, Charles G. Lang, Langvalley Farm, Glenarm, Md., who purchased three for $8,400, and John R. Kimberly, Neenah, Wis., who secured three for $4,200.

Buyers were located from Coast to Coast as the 28 breeders purchasing

Curtiss Candy Levity Hero, top bull that was purchased for $5,000 by William Y. Gilmore, Lake Geneva, Wis. In the background, Mr. & Mrs. Frank J. Mackey, Mukwonago, Wis., consignors; Mrs. Gerald M. Jenkins; William Y. Gilmore, and Gerald M. Jenkins, manager of Lookout Valley Farm.

nearly two-thirds of the arena for the breeders and buyers, as well as a complete set-up for serving the lunch. Immediately over the ring was a room used as the press box where representatives from the daily press and farm journals, as well as the clerk and force, were taken care of. Amplifiers were installed in the auctioneer's box so that everyone could hear distinctly.

Lunch at noon was the compliment of the Curtiss Candy Company and was served most expediently.

The sale was managed by Louis McL. Merryman & Sons with Mr. Merryman, reading pedigrees, E. M. Granger, Jr., as auctioneer, John B. Merryman, Louis McL. Merryman,

seed stock came from 15 different states. Thirteen animals made new homes in Wisconsin, 10 in Illinois, six in Indiana, five in Maryland, four in North Carolina, two each in Iowa and New York, and one each in Ohio, Virginia, Oregon, Michigan, Vermont, New Jersey, Massachusetts and Georgia. The consignors were located in 19 different states and also from Coast to Coast.

The arena was ideally arranged for the holding of this great sale. A portion was set aside and a half partition built with the ring and box attached to one side and the animals quartered in the other portion. This allowed

McDonald Farms Dairymaid that was purchased by Curtiss Candy Company for $8,000. In the background, J. M. McDonald, McDonald Farms, Cortland, N. Y., consignor; Delbert Kingston, Supt. of Curtiss Candy Company Farm; Mr. & Mrs. Otto Schnering, owners of Curtiss Candy Company Farm, and K. C. Sly, manager of McDonald Farms.

Inspecting the cattle before the sale.

Everett R. Beaty, Huntington, Ind.	
Meadow Lodge King Masterpiece (Bull)	2,250
Douglaston Baroness Florette	1,800
Charles Bolton, Lyndhurst, Ohio	
McDonald Farms King's Kella	13,500
Mrs. J. E. Carroll, Falmouth, Va.	
Douglaston Baroness Marjory	3,200
Curtiss Candy Company, Chicago, Ill.	
Lookout Valley Modern Mirth	1,000
McDonald Farms Dairymaid	8,000
St. James Champion Belle	1,200
Quail Roost Nomax Rose	3,700
Wey Acres Levity Piper (Bull)	1,000
Wey Acres Levity Kay	1,800
Hill Girt Phyllis	1,000

Western Glow Butterfat Maryann was flown from her home in Bow, Wash., via San Francisco to Chicago, by United Air Lines' Cargoliner. "Maryann" is giving nearly seven gallons of milk per day and during her brief stop at the Airport in San Francisco she gave 17.7 lbs. of milk. George W. Emde, in his official capacity as Executive Committee member of The American Guernsey Cattle Club, supervised the weighing of the cow's milk which was done by C. W. Robinson.

Western Glow Butterfat Maryann was consigned to the sale by Friedrichs Brothers and Ben may be seen holding her in the center picture.

Two days after these pictures were taken she topped the Curtiss Candy Sale, being purchased by M. C. Fleming, Troutdale, Ore., for $14,600.

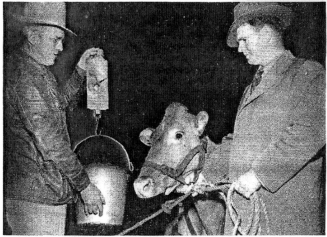

Jr., and Robert Seitz in the ring, Lizzie Merryman and Kitty Warfield running the slips and Helen Billingsley and Josie Merryman at the desk.

Following is the sales list:

Dr. W. H. Acker, Waterloo, Iowa	
Valleyland Levity Cecile	$2,300
Owendale's Bomber's Verna	1,100
O. P. Alford III., Queenstown, Md.	
Franchester Levity Ace (Bull)	1,950
Chester F. Allen, Kenosha, Wis.	
Langwater Maurader (Bull)	800
Walter H. Ball, Huntington, Ind.	
Raemelton Levity King (Bull)	1,100
Welcome-In Forward's Bonnette	2,050
Hydrangea of Chicona	1,700
Hilaria of Chicona	1,700
Afton's Golden Marie	10,000
M. C. Fleming, Troutdale, Ore.	
Western Glow Butterfat Maryann	14,600
William Y. Gilmore, Lake Geneva, Wis.	
Curtiss Candy Levity Hero (Bull)	5,000
V. L. Graf, Richmond, Mich.	
Starr Farm Miss Flo	1,000
Western Howland, Brattleboro, Vt.	
Meadow Lodge Gay Flute	4,600
John R. Kimberly, Neenah, Wis.	
Basquaerie Gay Rosalind	1,000
Skyline Triumph's Repose	2,200
Leahaven Maxim Star	1,000
Charles G. Lang, Glenarm, Md.	
Langwater Hornblend	1,700
St. James Philosopher's Hebe	1,300
Riegeldale Babette	5,400
R. A. Loeb, Clinton, N. J.	
Curtiss Candy Levity Barbara	3,600
John J. Mackin, Green Bay, Wis.	
Green Manor Daisetta	1,050
J. M. McDonald, Cortland, N. Y.	
Boulder Bridge Tulip	2,000
Western Glow Royal Levity	3,000

(Please turn to page 1299)

CARTER GUERNSEY
Sales......SERVICE......Advisory

You are invited to make use of my lifetime experience with Guernseys (Family Herd Est. 1912). To advise on and purchase for your Breeding Herd or Golden Guernsey Milk Production. Familiar with leading herds and many smaller ones.

Will Be Available at Eastern Sales

HENRY H. CARTER

Phone 742-W   Winding Lane Farm   ROCKVILLE, MD.

DANA'S NEW PREMIUM NECK CHAIN

Extra large, engraved numbers and letters • Aluminum alloy disc • Durable, long wearing • Length easily adjusted.

• Identification plate made from quarter-inch thick aluminum alloy — extra strong, long wearing, absolutely rust-proof. 3½" high, 2¼" wide, 1½" numbers. Engraved on both sides — numbers or letters, filled with black enamel. Chain is extra strong welded twist link design — lies flat — more comfortable to animal. Key ring type fasteners. 40" of chain supplied.

Quantity prices, F.O.B. Hyde Park, Vt.:
- Numbers 1 through 9 .... $1.40 each
- Numbers 10 through 99 ...... 1.65 each
- Numbers 100 through 999 .... 1.90 each

Remittance must accompany order

SEND FOR LATEST CATALOG of breeders' supplies, identification equipment, including many scarce items.

C. H. DANA CO., INC. (Est. 1861)
45 Main Street, Hyde Park, Vermont

GLENN LECKY
Auctioneer
of Pure Bred Guernseys
Holmesville - Ohio

For tomorrow's records feed MANAMAR today

This is another dam-daughter life record pair which has brought fame to Overbrook Dairy, Essex County, New Jersey, where ManAmar has been fed continuously since 1929. Combined Lifetime Total — 260,635 lbs. milk; 11,686 lbs. fat.

DAM Chesney Belle Sweet (V.G.)
DAUGHTER Essex Riverside Belle Sweet (V.G.)

We believe a feeding program including CATTLE ManAmar can help you get maximum results. ManAmar's sea-born proteins, vitamins and minerals supply "stabilizing" factors hard to get in any other way.

If Your Dealer Cannot Supply You, Write Dept. G

PHILIP R. PARK, INC.
618 S. Dearborn St., Chicago 5, Ill. or San Pedro, Calif.

Dunwalke Pride's Gem.............	180
Arch Lane Butterfat Fanny........	300
Stony Ford Douglas' Shelia.......	270
Arch Lane Butterfat Rose.........	180
L. B. Wescott, Clinton, N. J.	
Arch Lane Butterfat Belle......	160
Hilltop Butterfat Norma.........	650
Arch Lane Butterfat Pixie.......	600
Arch Lane Butterfat Mirth.......	150
Arch Lane Butterfat Norabelle...	275
Henry I. Winner, Mt. Holly, N. J.	
Jane of Rose Lane...............	275
John L. Winston, Gladstone, N. J.	
Arch Lane Butterfat Ellen........	350
Golden Rosalie of Rose Lane.....	400
Wood Ford Danna................	210
George Zwilgmeyer, Blairstown, N. J.	
Arch Lane Butterfat Faith........	190
Arch Lane Butterfat Clara........	320
Arch Lane Butterfat Lil..........	340

Curtiss Sale ...
(Continued from page 1269)

Louis McL. Merryman, Sparks, Md.	
Silver Forest Royal Valor (Bull)..	3,000
Sumner Pingree, South Hamilton, Mass.	
Howland Noble Juniper.........	1,600
Riegeldale Farm, Trion, Ga.	
Coronation Modiste.............	2,050
J. H. Rustman, Bassett, Wis.	
Pine Manor King's Siren........	950
George Schwemmer, Mukwonago, Wis.	
Curtiss Candy P. Etta...........	1,250
Shoal Falls Farms, Inc., Hendersonville, N. C.	
St. James Champion Julie.......	1,100
Riegeldale McK Becky..........	2,000
Cesor Maxim's Kitty............	1,000
Wandamere Hypatia............	1,100
W. B. Vilter, Hartland, Wis.	
Riegeldale Emory's Frances.....	2,000
Dorien's Lily..................	4,700
Elmer T. Voigt and Edware Klann, Chilton, Wis.	
Boulder Bridge Nestor (Bull)....	1,000
F. L. Weyenberg, Thiensville, Wis.	
Curtiss Candy Noble Deborah...	10,500
Orville J. Whitaker, Platteville, Wis.	
Liberty's Select................	900
Mark J. Woodhull, Angola, Ind.	
Hominy Hill Honeymoon.......	1,150
Churn Creek's Valor's Hilma....	1,500

Where to Buy GUERNSEYS
A Service to help both the Buyer and Seller

Rate: $3.50 per inch or 60 words
No proof submitted
All animals listed are for sale

FLORIDA

FOR SALE: Pure bred, registered Guernsey bulls out of high producing well-bred dams. Reasonable prices. Write for our latest bull list. V. C. Johnson, Dinsmore Farms, Dinsmore.

MISSOURI

FOR SALE: Herd Bull, Green Meads Justin, by Langwater County Squire and out of Green Meads Queen Jalna, 11214.9—614.0—GG, 15016.5—772.2—AA, by Langwater King of the Meads. Justin sired the Grand Champion Cow; Second Senior Yearling Heifer; 2nd Bull Calf, maternal sire; First Heifer Calf, Springfield, Missouri, Parish Show, 1946. Justin's first 3 daughters produced: Countess, 13997.0—632.0—F; Sebon, 9849.0—501.0—GHI (pending); Julia, 9268.0—449.0—D. H. I. A. age 2 years; all 2-time milking. Good breeder, sensible to handle. Age 7 years. Vaccinated. Price $2500. Will take good heifers as part payment. Parminter and Sons, Lockwood, Missouri.

NEW YORK

FOR SALE: Bull Calf, sired by a son of Langwater Grenadier, that has 42 A. R. daughters up to 690 lbs. fat. Calf's dam is a daughter of Langwater Chum that has 14 A. R. daughters up to 724 lbs. fat. Ask for pedigree and farmer to farmer price. V. M. Butterfield, Brockport.

OHIO

FOR SALE: Herd Sire, high record, partly proved, Borrodale Chief's Harlequin. First tested daughter, 10010.2—563.6—FF—twice-a-day milking. His Dam, 16342.4—585.9—AA, daughter of Imp. Sailor Lad V. of the Fontaines. Sire's Dam, 17668.2—893.5—A. Good individual, quiet, large type. Sure breeder. Accredited. Edward Witschi, Mendon.

Please mention the JOURNAL when answering our advertisers.

FRANKLIN Uterine Capsules
A valuable aid in removing afterbirth and treats septic infection. 3 capsules $1, 12 for $3, postpaid.

CATALOG FREE! **O. M. FRANKLIN SERUM COMPANY**
Livestock Exchange Building   DENVER, COLO.

FOR NOVEMBER 1, 1946

Western Glow Butterfat Maryann

Top of the 1946 Curtiss Candy Sale in Chicago selling for $14,600.00

l to r: M.C Flemming buyer, Helen Friedrichs and Ben Friedrichs representing Western Glow

*She must have been a very short cow as my Grandmother Helen was about 4'11" tall!

…Or they might have had Grandma standing on something.

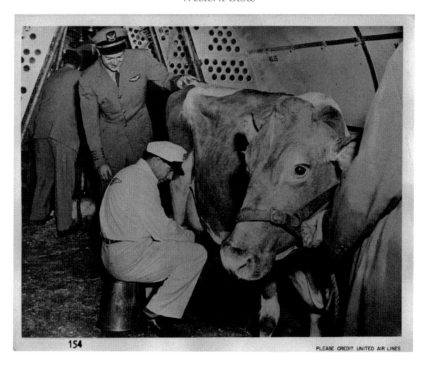

Maryann on the plane in Denver being milked on the way to Chicago.[3] These are press picture I found this summer on Ebay. There were articles in the Seattle papers and many others.

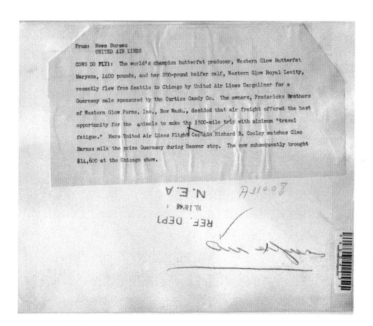

In 2015 dollars the sales price would be $174,000.00

---

[3] Credit to United Airlines New Bureau

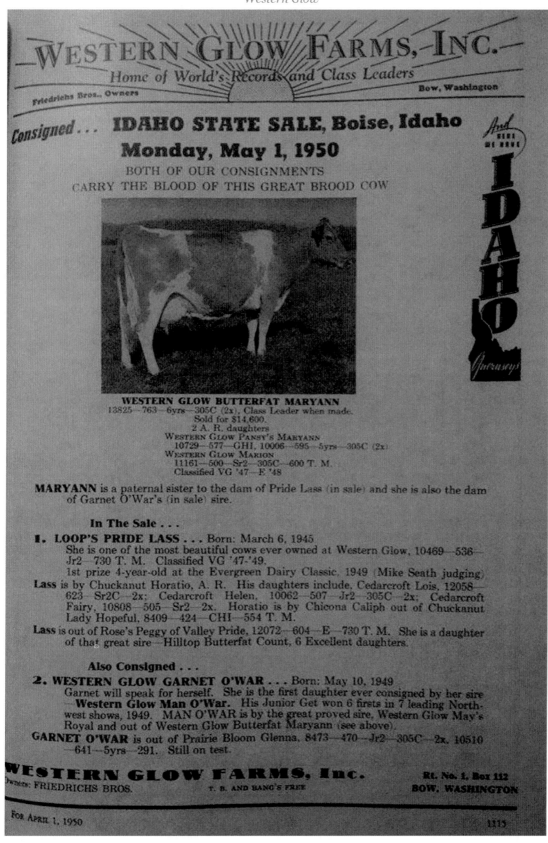

For the next several years the Maryann story was used in Western Glow ads. The Count blood was sought after in the early days of AI there were many bulls that traced back to him.

Chapter Four
The Influence of Virtue

Western Glow Fames Virtue

One Cow Three Sons

Fames Virtue is still a mystery to me. I heard her name repeatedly growing up though she was older that I was. She was out of **Western Glow Royal Fame**, a **Silver Forest May Royal** son out of one of the original daughters of **Hilltop Butterfat Count**. **Developer** was also in the pedigree of **Virtue of Aldarra** her dam. Royal Fame was used both at Western Glow and Boeing's Aldarra Farm. I could not find a picture her three sons but I am sure they exist as two of them were at Curtiss Breeders and one at ABS.

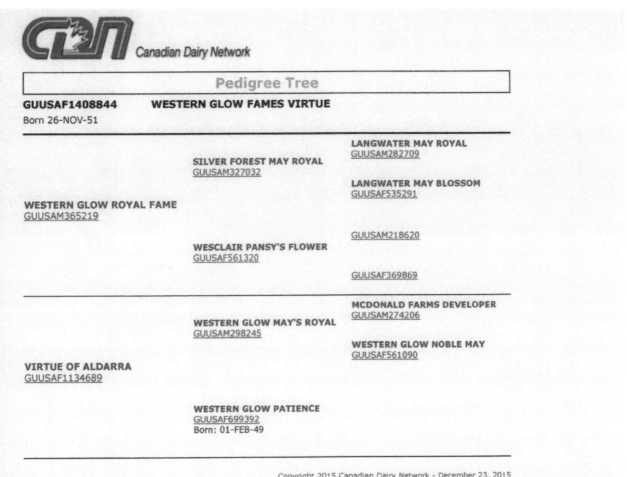

Her sons intertwined throughout the breed and by my count there were 34 All Americans in milking form that descended from hers sons and their offspring from 1962 with All American Aged cow Gayoso View Fames Debbie a Henslee Farms V Fame daughter. To the All American Aged cow in 1988 a daughter of Housley Dari Fayvor who also traces back to Henslee Farms V Fame on his dams side. It is amazing to me how far the bloodlines transmit in line bred herds. Without it the lines just disappear on the sires side. There is a current fear of inbreeding but history shows that done with intent it can be very successful.

The Sons

V Fame is found in the pedigree of many popular sires in Guernsey history. **Applebrook Fames Corporal** was a V Fame son that sired the great **Fox Run AFC Faye Ex 95** with great production and the dam of **Housleys Dari Fayvor**

Two full brothers **Top Command** and **Viscount** also sired several All Americans and Top Command was the sire of **Clovelly Top Hornet** another sire of All American offspring.

Canadian Dairy Network

Pedigree Tree

GUUSAM549235   **GAYOSO VIEW TOP COMMAND**
0040GU00179   Born 23-AUG-59

- **HENSLEE FARMS GALAHADS VIRGEL**
  GUUSAM512972
  Born: 08-DEC-53
    - **LANGWATER SIR GALAHAD**
      GUUSAM365829
      Born: 01-MAR-47
    - **VIRTUE OF ALDARRA**
      GUUSAF1134689
        - **WESTERN GLOW MAY'S ROYAL**
          GUUSAM298245
        - **WESTERN GLOW PATIENCE**
          GUUSAF699392
          Born: 01-FEB-49

- **WESTERN GLOW FAMES VIRTUE**
  GUUSAF1408844
  Born: 26-NOV-51
    - **WESTERN GLOW ROYAL FAME**
      GUUSAM365219
        - **SILVER FOREST MAY ROYAL**
          GUUSAM327032
        - **WESCLAIR PANSY'S FLOWER**
          GUUSAF561320
    - **VIRTUE OF ALDARRA**
      GUUSAF1134689
        - **WESTERN GLOW MAY'S ROYAL**
          GUUSAM298245
        - **WESTERN GLOW PATIENCE**
          GUUSAF699392
          Born: 01-FEB-49

Copyright 2015 Canadian Dairy Network - December 21, 2015

Canadian Dairy Network

Pedigree Tree

GUUSAM529456   **HENSLEE FARMS VISCOUNT**   **VISCOUNT**
0029GU00766   Born 07-JUL-56

- **HENSLEE FARMS GALAHADS VIRGEL**
  GUUSAM512972
  Born: 08-DEC-53
    - **LANGWATER SIR GALAHAD**
      GUUSAM365829
      Born: 01-MAR-47
    - **VIRTUE OF ALDARRA**
      GUUSAF1134689
        - **WESTERN GLOW MAY'S ROYAL**
          GUUSAM298245
        - **WESTERN GLOW PATIENCE**
          GUUSAF699392
          Born: 01-FEB-49

- **WESTERN GLOW FAMES VIRTUE**
  GUUSAF1408844
  Born: 26-NOV-51
    - **WESTERN GLOW ROYAL FAME**
      GUUSAM365219
        - **SILVER FOREST MAY ROYAL**
          GUUSAM327032
        - **WESCLAIR PANSY'S FLOWER**
          GUUSAF561320
    - **VIRTUE OF ALDARRA**
      GUUSAF1134689
        - **WESTERN GLOW MAY'S ROYAL**
          GUUSAM298245
        - **WESTERN GLOW PATIENCE**
          GUUSAF699392
          Born: 01-FEB-49

Copyright 2015 Canadian Dairy Network - December 21, 2015

Other sires from the V Fame bloodline

Clovelly Top Hornet

He was a **Top Command** son out of a **V Fam**e daughter. There are three crossed to **Virtue of Aldarra** in his pedigree and he also sired many All Americans and cows that really milked. He had three direct crosses to Virtue of Aldarra

This picture was reprinted from The Guernsey Breed - An Illustrated Chronicle by C.B. Harding. l to r: Helen Friedrichs, Floyd Krauter, EB Henslee, unknown, Doris Friedrichs, Jake Friedrichs, Ben Friedrichs, Jim Christan, and Ray Carr on the halter.

Virtue of Aldarra

She was the dam of **Western Glow Fames Virtue.** She was sired by **Western Glow Mays Royal** out of **Western Glow Patience.** She must have been sold in dam when Boeing purchased the young stock at Western Glow and then repurchased from Boeing. She was sold to Henslee Farms in 1952 for $5,600.00 a year after **Western Glow Fames Virtue** was born. She appears in the pedigree of **Clovelly Top Hornet** three times. Some of these picture are out of order but it has been a continual process of discovering these older pictures.

Chapter Five
The Duchess Family

A Different Branch of the Family Tree

The Duchess family at Western Glow brought all the pieces together of a lifetime of work. Maryann was like winning the Super Bowl and for many years the benefits continued as artificial breeding units sought the bloodlines of the foundations sires and their daughters for sires of sons. Digging through a hundred pedigrees on the CDN website the origin of the Duchess family comes back to an own daughter of Hilltop Butterfat Count, her name was **Western Glow Butterfat Duchess** she was bred to **Western Glow Marymost** a son of **Western Glow Treasure** a maternal brother to another of the foundation Count daughters **Wesclair Pansy's Flower**. The resulting mating produced **Western Glow Merry Duchess** who was bred to **Fame of Aldarra** a **Western Glow Royal Fame** son that on his sires side went back to **Wesclair Pansy's Flower** again. The resulting son of **Merry Duchess, Western Glow Butterfat Traveler 21GU115** (his 12/14 sire summary shows him +7.5 on DPR) was the sire of **Western Glow Travelers Duchess.**

Western Glow Travelers Duchess

Pictured above. She was born October 13, 1955. She classified VG 86 at 12-11 and made at 8-09 2x 317d 11,980 6.1% 727 she calved 9 times and her daughters and sons would cause quite a stir across the country. She had 3 Excellent daughters and 2 out of her 3 sons went into A.I. When doing the research for this book it showed two of her dams daughters sold to Canada, one at Maple Springs Farm produced several offspring, also her grand dam was sold to Henslee Farms. I remembered her as an old cow that always was talked about. Growing up we were surrounded by the Duchess family. She had an amazing impact through her offspring.

Western Glow Fond Gay Duchess Ex-94

The oldest daughter of **Travelers Duchess** was sired by Fritzlyn Jeanette's Flash. She made 3 records over 900 fat, her best records was   7-07 2x 360 18,890 5.3% 1008 F
She made 116,896 M 5.4% 6276 F Lifetime

All American 4yr old 1963   All American Aged Cow 1964   Res. All American Aged Cow 1965

She sold for $10,000 cash in 1963 to Woodacres in Princeton, New Jersey and resold in 1966 for $18,500 to Walker Farm in Mississippi. The picture on the front of the book was taken in August of 1963 the day Fond Gay left by train for New Jersey to go to her new home at Woodacres. Grandpa Ben and John Rietman accompanied her on the trip.

WESTERN GLOW FOND GAY DUCHESS

AMERICAN GUERNSEY ASSOCIATION

```
WESTERN GLOW FOND GAY DUCHESS    USA000001920224                                              Current Date: 1-7-2015
    Born: 01-05-1959                        WESTERN GLOW M B DUCHESS         87 @ 9-02
 Breeder: WESTERN GLOW FARMS INC            7-04 308D 2X  13630M 6.0%  814F                    DHIA
          BOW              WA 98232         WESTERN GLOW LADY MARY DUCHESS   84 @11-00
   Owner: ANDERSON EDWARD C EST             8-00 306D 2X  15410M 5.5%  844F                    DHIA
          BUSHY PARK FARM
          P O BOX 98                              AGED                1964 AA
          WAKE             VA 23176              4 YRS                1963 AA
Appraisal:  94 @  8-10                            AGED                1965 RES AA
 2-02 305D 2X  13296M 4.9%    645F    DHIR       S & G CH             1964 N G S
      309D 2X  13432M 4.9%    653F    DHIA   1st AGED                 1964 N G S
 3-02 365D 2X  15091M 5.5%    825F    DHIA   2nd AGED                 1965 N G S
 4-05 305D 2X  17305M*5.5%    958F*   DHIR   1st 4 YRS                1963 N G S
 5-05 305D 2X  10937M 6.0%    656F    DHIR       BEST UDD CH          1964 N G S
      314D 2X  11118M 6.0%    668F    DHIA   2nd BEST UDD 4 YRS       1963 N G S
 6-05 305D 2X  14345M 5.1%    734F    DHIR   1st BEST UDD AGED        1964 N G S
      309D 2X  14477M 5.1%    742F    DHIA       S & G CH             1963 N J
 7-07 305D 2X  17180M 5.3%    916F    DHIR       S & G CH             1964 E GUERNSEY
      360D 2X  18890M 5.3%   1008F    DHIA       S & G CH             1965 N J
 8-08 305D 2X  16900M 5.1%    858F    DHIR   1st AGED                 1964 E GUERNSEY
      2662D   116869M 5.4%   6276F  0.0%   0P   LIFE  1st AGED        1965 N J
                                             1st 4 YRS                1963 N J
                                             2nd PRODUCE              1965 W WASH

                                             FRITZLYN FLASH                  USA000000411629
                                             -----------------------------------------------
FRITZLYN JEANETTES FLASH     USA000000475548
------------------------------------------------FRITZLYN GYPSY KINGS JEANETTE  USA000000969573
12/14 USDA PTA  -2316M +0.21%  -74F +0.02%  -73P
  +4.6DPR -1.2PL  -407NM$   0NM%ile  -400CM$   95% Rel
   AGA 12/14  PTAT -3.4 -1.1UDC -2.0FLC 63%Rel  PTI -298

 ST   SR   BD   DF   RA   TW   RV   RL   FA
-6.1 -4.2 -3.8 -3.7 L1.7 -3.7      0.0 L1.4
 FU   RH   RW   UC   UD   TP   TL
-0.9 -2.4 -2.6 -1.2 S0.1 W0.8 -1.2               WESTERN GLOW BUTTERFAT TRAVLER USA000000511205
                                                 ----------------------------------------------
WESTERN GLOW TRAVLERS DUCHESS  USA000001682143   12/14 USDA PTA  -2616M +0.13%  -100F +0.10%  -71P
------------------------------------------------   +7.5DPR -2.6PL  -546NM$  0NM%ile  -524CM$  79% Rel
Appraisal:  86 @ 12-11
 2-02 305D 2X  10081M 5.3%    535F    DHIR
 3-02 303D 2X  10982M 5.7%    628F    DHIR
 4-02 305D 2X   9911M 5.2%    520F    DHIR
 5-05 326D 2X  11563M 5.2%    602F    DHIA
 6-06 311D 2X  10588M 5.3%    560F    DHIA
 7-06 300D 2X   9780M 5.0%    489F    DHIR
 8-09 317D 2X  11980M 6.1%    727F    DHIA
10-01 293D 2X  12010M 5.7%    690F    DHIR
11-03 279D 2X  11120M 5.3%    594F    DHIR
WESTERN GLOW TRAVELERS DUANN       91 @ 5-09
 5-00 305D 2X  13370M 5.2%    692F    DHIR
WESTERN GLOW NOBLE GAY DUCHESS     90 @ 7-01
 5-05 310D 2X  16010M 5.3%    854F    DHIA
WESTERN GLOW NOBLE DUTCH GIRL      83 @ 7-04
 5-02 315D 2X  16350M 5.5%    905F    DHIA
```

Of her three sons in A.I. **Walker Farm Grand Duke** left some great cows in the Western Glow herd, a granddaughter of one would also be an All American and one of the best cows bred at Western Glow. Fond Gay's daughters also were great brood cows in their own right. The family contributed fertility, longevity, and exceptional butterfat test. Notice the consistent calving intervals on both Fond Gay and her dam.

WALKER FARMS GRAND DUKE
Born August 17, 1966

We are happy to have negotiated the sale of this bull from Walker Farms to ABS. His great total inheritance makes Grand Duke truly a "Breeders' Bull". He carries a commingling of the very best in the Guernsey breed through the planning and work of a number of foremost breeders. He carries a double cross to the great Butterfat Duchess and Maryann families. We have known the family for many generations and it is noted for the outstanding animals that stem from this proved source.

Canadian Dairy Network

Pedigree Tree

GUUSAM573574	WALKER FARMS GRAND DUKE	GRAND DUKE
0029GU00775		Born 17-AUG-66

		FRITZLYN FLASH
	FRITZLYN JEANETTES FLASH	GUUSAM411629
	GUUSAM475548	
	Born: 28-DEC-50	
		GUUSAF969573
WESTERN GLOW DARIMOST		
GUUSAM546331		
Born: 14-MAY-59		
	RIVER ROAD ROYALS DEBBIE	
	GUUSAF1146132	

		FRITZLYN FLASH
	FRITZLYN JEANETTES FLASH	GUUSAM411629
	GUUSAM475548	
	Born: 28-DEC-50	
		GUUSAF969573
WESTERN GLOW FOND GAY DUCHESS		WESTERN GLOW BUTTERFAT TRAVLER
GUUSAF1920224		GUUSAM511205
Born: 05-JAN-59		Born: 27-OCT-53
	WESTERN GLOW TRAVLER'S DUCHESS	
	GUUSAF1682143	
	Born: 13-OCT-55	GUUSAF1256439
		Born: 28-FEB-50

Copyright 2015 Canadian Dairy Network - December 29, 2015

Pictured above is **Walker Farm Grand Duke 29GU775** he was at ABS and was a **Western Glow Darimost** son out of **Western Glow Fond Gay Duchess**. He was truly a breeders bull with both sire and dam sired by **Fritzlyn Jeanette's Flash** with the blood of Butterfat Maryann on the top side and the Duchess family on the bottom side. One of his daughters **Western Glow Dukes Beverly** VG 87 and over 983 of fat and 111,980 milk lifetime with 5.7% and 6343 of fat and she was the dam of **Western Glow P N Beloved** at Ex 93 and became the 1977 All American 4 year old. More on her later.

Another son of Fond Gay that went into A.I. was **Walker Farms Gay Rebel 40GU196** he was a Top Command son that brought both **Fames Virtue** and the Duchess family together in one package.

The Daughters of Fond Gay Duchess

Western Glow M.B. Duchess VG 87 @ 9-02

M.B Duchess was sired by **Western Glow Muriels Bonanza** he was out of **Western Glow F Muriel** a daughter of **Coldsprings BR Forecaster** (He also sired Henslee Farm V Fame) **Muriel** was also the 3rd dam to **Western Glow Melbas Champion 11GU208** who was at Carnation Genetics.

MB Duchess did well in the show ring was VG 87 and made 7-08 308d 2x
13630 6.0% 814

Her son **Western Glow May Rose Duke 72GU36** was widely used in Canada with over 450 daughters, he was sired by **Lyrene May Rose Prince.**

Her son **Western Glow MB Duke** was used at home and showed in the Northwest and Canada. His pedigree had Fond Gay top and bottom. He was intensely inbred and would have been an interesting bull to use in an outcross herd. He shows up deep in the pedigree of **7GU302 Rozelyn Pat Mar Goliath ET** through the paternal side of his 3rd dam.

WESTERN GLOW M B DUKE

Summary | Progeny | Pedigree | Owner | Inbreeding

Breed Association

Pedigree Tree

GUUSAM577577 **WESTERN GLOW M B DUKE**

Born 04-AUG-68   15.92%INB   3%R

WALKER FARMS GRAND DUKE
GUUSAM573574
Born: 17-AUG-66

- **WESTERN GLOW DARIMOST**
  GUUSAM546331
  Born: 14-MAY-59
  - **FRITZLYN JEANETTES FLASH**
    GUUSAM475548
    Born: 28-DEC-50
  - **RIVER ROAD ROYALS DEBBIE**
    GUUSAF1146132

- **WESTERN GLOW FOND GAY DUCHESS**
  GUUSAF1920224
  Born: 05-JAN-59
  - **FRITZLYN JEANETTES FLASH**
    GUUSAM475548
    Born: 28-DEC-50
  - **WESTERN GLOW TRAVLER'S DUCHESS**
    GUUSAF1682143
    Born: 13-OCT-55

WESTERN GLOW M B DUCHESS
GUUSAF2095437
Born: 26-MAR-61

- **WESTERN GLOW MURIELS BONANZA**
  GUUSAM546330
  Born: 11-APR-59
  - **NYALA ERMINES BONANZA**
    GUUSAM496783
    Born: 26-MAR-52
  - **WESTERN GLOW F MURIEL**
    GUUSAF1419256
    Born: 02-OCT-51

- **WESTERN GLOW FOND GAY DUCHESS**
  GUUSAF1920224
  Born: 05-JAN-59
  - **FRITZLYN JEANETTES FLASH**
    GUUSAM475548
    Born: 28-DEC-50
  - **WESTERN GLOW TRAVLER'S DUCHESS**
    GUUSAF1682143
    Born: 13-OCT-55

Copyright 2014 Canadian Dairy Network - December 31, 2014
Privacy | Disclaimer | Sitemap | Contact Us | Français

Other daughters of Fond Gay Duchess:

Western Glow Lady Gay Duchess sold to Lake Louise Farm. Sired by **McDonald Farms Jolly Val**

Also Fond Gay was the dam of **Western Glow Lady Mary Duchess**

VG 84

8-00 306d 2x 15,410 5.5% 844 she was also sired by Jolly Val and had a Darimost daughter that was the dam of **Western Glow Dari Financier** a bull that was shown extensively and was nominated for All American.

The characteristics of the Duchess family were cows that lasted and cows that exceeded 5.0% butterfat. These cows bred back every year and were intensely line bred. This was before embryo transfer so the only way you could sell females was to develop families that produced heifer calves. Over time you had more family members to choose from for breeding and marketing. The key to longevity is also care and cow comfort. With Jake taking care of the feeding and milking of the milking herd the care was extraordinary. In later years Ben Jr. took over the farming duties and feeding as well as doing most of the artificial breeding.

The Duchess family put their mark on the 1960's at Western Glow putting several bulls in A.I. and building the base of females to be marketed.

Other Daughters of Travelers Duchess

Western Glow Travelers Duann Ex 91 she was sired by **Western Glow Maryann's Traveler** (a maternal brother to **Western Glow F Muriel**) by **Western Glow Butterfat Master.** She left one son by **Fritzlyn Jeanette's Flash** that saw limited use. This picture is one of my favorites. She is the kind of cow i would love to flush today.

Western Glow Noble Gay Duchess Ex 90 she was sired by **Norgerts Royal Nance** and made 5-05 310d 2x 16,010 5.3% 854 she had a son by **Housleys J Champion** that went to ABS **29GU787 Western Glow Gay Champion.**

Chapter Six
The Story of Darimost
6,666 daughters in 1,120 herds

180　　　**WESTERN GLOW DARIMOST**　　　546331

Born May 14, 1959　　　Guernsey
Bred by Western Glow Farms, Inc., Bow, Washington
Weight 1920
Owned by All West Breeders

All West Breeders Preliminary Summary
　　　　　　　　　　　　　　　(305d 2x M.E.)
　　　　　　　　　　　　　　Milk　　%　　Fat
16 A.I. daughters average　　11,372　4.66　530
Plus Herdmates　　　　　　　+762　　　　+40

Official Guernsey Type Proof
　7 classified daughters　83.4%

Characteristics of Daughters
　Improves both milk and butterfat production
　Improves size, height, and stretch
　Improves speed of milkout and quality of udder
　Dispositions highly desirable

No other bull in the history of Western Glow saw as much use as **Western Glow Darimost 13GU180** he was a Gold Medal sire sired by **Fritzlyn Jeanette's Flash** the last sire brought into the program with a concentration of Langwater blood. Darimost's 4/15 proof shows 6,666 daughters in 1,120 herds he also is +6.5 on DPR and 2.47 Somatic Cell Score.

His daughters earned 11 All American awards in milking form spanning from 1964 to 1980

His Dam **River Road Royals Debbie** Ex 91 made her last record at

15 years 6 months 294d 2x 11940 5.2% 625

Her sire was **Western Glow Maryann's Royal** the **Silver Forest May Royal** son of **Western Glow Butterfat Maryann** the **Hilltop Butterfat Count** daughter who topped the 1946 Curtiss Candy Sale in Chicago. The Count's on the bottom side and **Fritzlyn Jeanette's Flash** and the last shot of Langwater blood on the top side. He was richly bred and sired a pattern through his daughters and sons. There were years when the Holstein line up at All West Breeders was struggling Archie Nelson told me "Darimost kept the lights on, we sold a lot of semen out of him in those years."

Cleverlands Darimost Chrystal Ex 95

All American 3 year old 1975

All American Aged Cow 1979

All American Aged Cow 1980

All American Awards for Darimost in Milking Form 1964-1980

Year	Cow	Age
1964	Roetta Dariann	2 yr old
1968	Candy Jewel Ex 92	Aged Cow
1970	Western Oaks D Velita Ex 92	3 yr old
1971	Western Oaks D Velita Ex 92	4 yr old
1974	Will O West Dari Edean Ex92	4 yr old
1975	Cleverlands Darimost Chrystal Ex95	3 yr old
1976	Will O West Dari Edean Ex92	Aged Cow
1977	Golden Acres Dari Lillian Ex 94	Aged Cow
1978	Circle View Darimost Maisy Ex92	Aged Cow
1979	Cleverlands Darimost Chrystal Ex 95	Aged Cow
1980	Cleverlands Darimost Chrystal Ex 95	Aged Cow

The **Darimost** daughters matured well and held up, many became bull mothers. He also had several key sons. Another sire of many All Americans was **Housleys Dari Fayvor 29GU784** his dam **AFC Faye Ex 95** was All American Aged Cow in 1969 and traced back to **Henslee Farms V Fame** on her sires side.

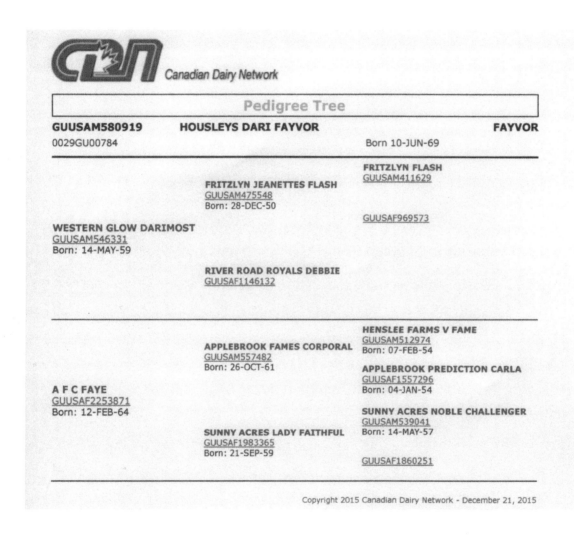

Darimost was used heavily at Western Glow with 1/3 of the herd consisting of his milking daughters in the late 60's and early 70's.

There were select Darimost daughters purchased by Western Glow as well:

Pathway D Ellen Ex 90

bred by Dick Dudonsky

One key purchased daughter was from Cleverlands Farm in California.

Cleverlands Darimost Alice Ex 90 was a silky hided cow that was extremely milky making over 24,000 pounds of milk in the 1970's. Her dam was sired by **Western Glow Duchess Superior** a full brother to Fond Gay Duchess so she had two close crosses to **Fritzlyn Jeanette's Flash**. Her son **Western Glow TH Alstar 72GU59** a Top Hornet son was widely used in Canada with over 1600 daughters on his proof.

He sired **Queenholm Alstar Ginger** Ex 90 she was 3x All American as a 2yr old in 1984, 3yr old in 1985, and 4yr old in 1986.

Alice also had a Hollirex son **Western Glow HR American 11GU218** that was sampled in the US.

Western Glow Duchess Superior sired the maternal grand dam of TH Alstar. He was the Flash full brother to Western Glow Fond Gay Duchess.

Western Glow Darimost GM his page in the 1969 All West Breeders Sire Directory

2180 WESTERN GLOW DARIMOST 546331 GOLD STAR

USDA: (5/69) Repeatability 94%
PREDICTED DIFFERENCE +542 MILK +32 FAT
495 Daughters 143 Herds 11,082M 4.8% 526F
AGCC: (1/69) 274 Classified Daughters Average 84.4
All-American Aged Cow, 1968
All-American Heifer Calf, 1968
Res. All-American Jr. Get, 1968

aAa — SRR-5

Born: 5-14-59 Weight: 2000

Upstanding	Body Capacity		Dairy Character	Mammary System			Feet & Legs	Rumps
	Width	Depth		Fore	Rear	Quality		
*	*	*	***	***	**	***	*	**

SIRE: 575548
FRITZLYN JEANETTE'S FLASH, GS

USDA: (1/69) Repeatability 94%
Predicted Difference +378M +27F
3101 Daus. 206 Herds 10,745M 4.9% 523F
AGCC: 473 Class. Daus. Avg. 85.1
40 Excellent Daughters

FRITZLYN FLASH
27 Daus. Avg. 9,030M 5.0% 453F
7 Class. Daus. Avg. 82.4
3 VG, 2D, 1F
$1700 Fritzlyn Dispersal, 1951

FRITZLYN GYPSY KING'S JEANETTE
Jr. 2 365d 13,945M 4.3% 601F
Sr. 4 365d 19,672M 4.5% 894F
$3500 Fritzlyn Dispersal, 1951

WESTERN GLOW MARYANN'S ROYAL
37 Daus. Avg. 9,156M 4.7% 466F
1 Over 800, 4/700, 8/600, 17/500
34 Class. Daus. Avg. 83.4
2 Ex., 12 VG, 14D, 5A

DAM:
RIVER ROAD ROYAL'S DEBBIE, EX-4X

7y	305d	10,739M	628F
15y	294d	11,940M	625F
6y	305d	10,550M	604F
13y	243d	10,297M	601F
10y	305d	11,243M	579F
8y	305d	10,320M	573F
14y	282d	10,798M	503F

11 Lact. Total: 121,168M 5897F

RIVER ROAD DARLENE
Sr. 2 305d 9,446M 4.6% 433F
Sr. 3 305d 8,556M 5.3% 453F
4 yrs. 305d 10,498M 5.2% 545F

Daughter
Candy Jewel, EX-91.5
All-American Aged Cow, 1968

Daughter
Roetta Dariann, EX
All-American 2 yr. old, 1964

Chapter Seven
Sun Glow Bev
by Lake Louise Patsy's Prince

Sun Glow Bev Ex 92

Most kids have other things pinned up on the wall of their bedrooms, I had Sun Glow Bev. I don't know the full story of how she came to Western Glow…I just knew l liked this strong speckled cow and her daughter.

She was sired by **Lake Louise Patsy's Prince** a son of **Lake Louise Royal Superb** who in turn was a son of **Western Glow Lou's Royal (left)** a son of **Western Glow Mays Royal** a **Developer** son out of **Western Glow Butterfat Lou** an own daughter of **Hilltop Butterfat Count.**

The story of **Sun Glow Bev** will always be tied to two other **Lake Louise Patsy's Prince** daughters **J.L.Princess,** and **Firview Prince's Mary Jo.** I remember standing at in the barn at the Calgary Stampede as a fellow breeder complimented the show string we had there. The best cow in the string was a great uddered Melbas Champion daughter named Mardey. Grandpa said "Thanks for the compliment considering I just sold the three best cows we had to Woodacres." Hobler and Mert Sowerby were looking to fill out their **Lake Louise Patsy's Prince** Get of Sire and he hit the jackpot with those three. I never did hear what he got for Mary Jo but I did over hear conversations that he got $6,500 a piece for Bev and J.L.

We did have a heifer on the ground out of Bev when she left; **Western Glow Dukes Beverly** she was sired by **Walker Farm Grand Duke** the **Darimost** son of **Western Glow Fond Gay Duchess.** Not a bad way to nail down a foundation. Beverly would go on to have 7 daughters 2 Excellent's and 3 Very Good. She herself would go VG 87 at 9-03 and over 983 of fat and 111,980 milk lifetime with 5.7% and 6343 of fat. She calved like clock work from 2-01 to 9-05. Her DPR on an older pedigree was +4.5

Dukes Beverly Daughters:

Western Glow LF Bev	VG 84	by Lyrene Financier
Western Glow PN Betsy	D 76	by Cedarbrook Polly's Nance
Western Glow PN Beloved	Ex 93	by Cedarbrook Polly's Nance
Western Glow NA Belva	VG 85	by Rosewood Nona's Ambassador
Western Glow BD Ballet	VG 80	by Western Glow Dari Brilliant
Markwell Fayvor Believe	Ex 90	by Housleys Dari Fayvor
Markwell Fayvor Bountiful		by Housleys Dari Fayvor

Of all these **Western Glow PN Beloved** developed into a once in a lifetime individual she was pictured nearly every year, she grew up tall and she grew up right. She was special from the start in 1973 she was 1st place Sr Calf at Puyallup. 1974 she was 1st Sr Yearling and Junior Champion at the NGS in Puyallup and Reserve All American Sr Yearling. She won the 1975 2yr old class at the Evergreen Classic in Mt Vernon, Washington. She won the 2 yr old class that year at the National Guernsey Show in Fresno and was Senior and Grand Champion rare for a 2 yr old, and Reserve All American 2 yr old. In 1976 she was 1st 3 yr old and Best Udder at the NGS in Puyallup and won Senior and Grand Champion again as well. She was Honorable Mention All American 3 yr old in 1976. In 1977 she won her class at the NGS in Fresno for the 3rd year in a row. She was All American 4 year old in 1977.

She was sold to Neal Jacobs of Copper State Dairy in Arizona. She sold for $7,500.00 and later scored Ex 93.

Wester Glow PN Beloved as a Sr Yearling

Western Glow P N Beloved 2 yr old

Western Glow PN Beloved 3 yr old

Western Glow PN Beloved 4y r old

She ended up at William Askews in Malin, Oregon. Beloved had two daughters at Western Glow, **Western Glow NA Ballerina** Ex 90 7-04 345 2x 14,560 5.3% 770 by **Rosewood Nona's Ambassador** and **Western Glow DF Be My Love** D 79 3-06 12,940 5.4% 697 by **Western Glow Dari Financier**. Another daughter was **Askew Farms TH Beloved** VG 89 5-11 365 2x 14,890 4.4% 657F 3.5% 518P

Western Glow PN Beloved Ex 93 4-09 330 2x 19,350 4.4% 849

She was all you could ask for.

The crux of the story of **Sun Glow Bev** was that there was a chance to buy her back at the Woodacres Dispersal a few years later. Instead a group of which Western Glow was a part of bought back **J.L. Princess** instead…next chapter.

Sun Glow Bev was purchased by Alton Gunby of Maple Valley Farm in Burlington, Ontario in the Woodacres Dispersal, where she left several offspring. He got the buy of the sale as Sun Glow Bev would be one of the great brood cows at Western Glow…though she was there only a few short years.

Chapter Eight
And there were others....
J.L Princess Ex92
The other Lake Louise Patsy's Prince daughter

Another cow that was found as a diamond in the rough was J.L Princess. She was a rugged framed cow with a rump like a box car and a muzzle to match. she had injured her leg being loaded as a young cow and always had kind of a club foot that took away from her otherwise powerful presentation. She was a Patsy's Prince daughter out of a dam that went back to **Coldsprings BR Forecaster** on her sires side and **Flying Horse Bandmaster** on her dams side.

She sold to Woodacres in the late 1960's with Bev and Mary Jo. In the Woodacres Dispersal she was brought back to Western Glow. The decision to purchase her again was probably based on the A.I. potential interest rather than strictly a breeding program decision.

She had three sons in A.I. **Western Glow Wistar Premier 7GU1036** was the most widely used with over 500 daughters.

```
                        WESTERN GLOW WISTAR PREMIER
WESTERN GLOW WISTAR PREMIER    USA000000585495  7GU1036            Current Date: 3-16-2010
      Born: 08-09-1972  Tattoo LE W610
   Breeder: WESTERN GLOW FARMS INC FRIEDRICHS BEN JR & VAN
            BOW                   WA 98232
     Owner: SELECT SIRES INC
            BLAINE CROSSER
            11740 U S 42 N
            PLAIN CITY       OH 43064
01/10 USDA PTA  -1715M +0.19%  -47F +0.13%  -35P
  +4.0DPR +1.2PL  -106NM$    12NM%ile  -44CM$    99% Rel
  AGA 01/10  PTAT -2.6 -1.9UDC +0.0FLC 99%Rel  PTI -166

    ST    SR    BD    DF    RA    TW    RV    RL    FA
  -2.6  -1.7  -3.3  -3.8  L0.3  -1.9  S0.9  P0.4  S0.7         LYRENE CH WISTAR              USA000000470256
    FU    RH    RW    UC    UD    TP    TL
  -2.1  -3.6  -3.0  -1.2  S0.1  W2.3  -0.1

LYRENE MYRAS WISTAR            USA000000537680
                                                               -LYRENE FAVORITE MYRA         USA000001540439
01/10 USDA PTA  -2178M +0.16%  -72F +0.11%  -54P
  +2.1DPR -0.8PL  -305NM$   0NM%ile  -265CM$    96% Rel        Appraisal: 90 @ 8-10
  AGA 01/10  PTAT -2.6 -1.6UDC +0.1FLC 91%Rel  PTI -235         6-03  311D 2X  13190M 5.9%   775F        DHIA

    ST    SR    BD    DF    RA    TW    RV    RL    FA
  -2.7  -1.9  -2.8  -3.1  L1.0  -2.0        P0.6  L0.1
    FU    RH    RW    UC    UD    TP    TL
  -1.3  -3.5  -3.3  -1.3  S0.6  W2.1  +0.6                      LAKE LOUISE PATSYS PRINCE    USA000000549336  13GU185

                                                               01/10 USDA PTA  -2669M +0.16%  -94F +0.09%  -74P
J L PRINCESS                   USA000002215471                   +3.0DPR -0.7PL  -436NM$   0NM%ile  -420CM$  91% Rel

Appraisal: 92 @ 11-07
 4-07  332D 2X  16290M 5.1%   828F        DHIA                 SKY VALLEY LB JUNO            USA000001994293
 5-08  363D 2X  19770M 4.4%   879F        DHIA
 6-11  328D 2X  15580M 4.7%   740F        DHIA                 Appraisal: 75 @ 0-00
 8-00  351D 2X  18140M 4.7%   860F        DHIA
 9-03  308D 2X  13570M 4.8%   657F        DHIA
10-03  305D 2X  16000M 4.8%   760F        DHIR
11-06  305D 2X  16280M 4.5%   725F        DHIR
12-09  335D 2X  16300M 4.6%   749F        DHIA
       2701D    134480M 4.7%  6325F 0.0%   0P   LIFE
WOODACRES FOND PRINCE                        PTI  -261
D PRINCESS LORI                              90 @ 7-10
 5-00  323D 2X  16280M 5.1%   823F        DHIA
WESTERN GLOW DARI PATRICE                    90 @ 8-04
 3-01  319D 2X  16330M 4.8%   781F        DHIA
WESTERN GLOW DARI PRECISE                    82 @ 7-03
 4-02  332D 2X  14610M 5.2%   767F        DHIA
WESTERN GLOW FOND PRINCESS                   78 @ 3-06
 3-01  318D 2X  14810M 5.1%   762F        DHIA
```

He was sired by **Lyrene Myra's Wistar** a bull that could ruin a rump on a Guernsey cow like no other bull...and he made short work of J.L. Princess's box car rump.

Woodacres Fond Prince 7GU 1260 was a **FritzLyn Jeanette's Flash** son that saw limited use with 70 daughters in the CDN database. Flash was probably a little too smooth for J.L Princess and they were not dairy enough to milk well. **Western Glow PN Predictor 1GU127** was a young sire that did not make the line up as the mid 1970's brought a push for higher milk bulls.

J.L. made 134,480 lifetime with 6325 of fat with a 4.7% test this was without her 1st two lactations as she was not on test then. The daughters of J.L were tall and to the extreme. Two by Darimost of which **Precise** was tall and rugged and the other **Patrice** extreme dairy. A third daughter **D Princess Lori** Ex 90 by **Bay Meadows Billy's Don** that was purchased by Far West Farms was a powerhouse though she lacked openness and dairy character.

Western Glow Dari Precise VG 82

Her son Western Glow Financier Prince saw use at Western Glow

Western Glow Dari Patrice Ex 90

The Third Patsy's Prince daughter....

Firview Prince's Mary Jo was the third cow in the original Woodacres deal that sent the three Patsy's Prince daughters East. When she was milked off she looked good but she was a smoothy in this picture.

Western Glow Brilliant Mae
Beautifully uddered and balanced

Brilliant Mae in her wedding clothes was something to see, she was balance defined and when she was fresh and in bloom she was hard to get around in the show ring. She was a Liseter Brilliant daughter out of a Western Glow Melbas Champion daughter out of a Nyala Ermines Bonanza daughter. Brilliant Mae also had a beautiful Cedarbrook Polly's Emory daughter that had an udder like her dam.

Western Glow Brilliant Mae

Western Glow Emory's Monica

Western Glow Champions Mardey VG86

Mardey was the queen of the Prairie circuit for many years she had a beautiful udder that milked out like a dish rag. She won the Interbreed Best Udder Class on several occasions. Grandpa would milk her by hand as the other cows were being milked out ringside with machines. He would sit down and fill the large galvanized pail faster than the guys milking by machine. You should have seen him smile as he walked by the judge with the full pail. She was a joy to be around.

Chapter Nine
The Golden Years
The Washington State PDCA Hall Of Fame Award

l to r: Jake Friedrichs, Doris Friedrichs, Helen Friedrichs, Ben Friedrichs, Lawrence Newman

There were awards they received from the PDCA Hall of Fame to Master Breeder awards in the US and Canada that were greatly appreciated. The team of Ben and Jake made it possible. One without the other could never have achieved what they did together. The years had been good to Ben and Jake, they worked hard through good times and bad. And occasionally there would be arguments but they always stayed in the rut, always took care of the cows, always made sure people who came to the farm saw clean cows that were well fed.

We learned a lot of what was important in taking care of cows. Most importantly keeping them clean and well fed. While the cows were being milked the piles in the sheds were being picked up and the bedding leveled. The mangers were stocked with fresh hay and silage and when the cows came back out of the barn they were to be left alone so they could eat, rest, and chew their cud. Grandpa always said "the only time they are making money is when they are laying down chewing their cud." The cows always came first. Jake took care of the cows day to day and Ben took care of the calves and heifers. I can only remember one calf that died while we were growing up. It just happened to be the last calf of Traveler's Duchess.

It was hard work and long hours but those sunsets at Western Glow made it all worth it. Grandpa was always working the breeding program, whether it was finding a new female to work with or selecting a bull to use. He would line out a pedigree diagram and plug in the potential mating to see how it fit the program. He could rattle off cow families while on the phone in his office. You could learn a lot if you just listened.

You never knew who was going to show up…I remember we were cleaning heifer pens one Saturday and in the barn walks Pete Heffering and David Younger looking for sale consignments for a sale back East. We were notorious about getting cattle broke to lead, most of them got a halter on the first time they got on the truck to go to a show or sale. We worked with heifers the night before the show and tugged on them more than led them. We clipped udders and tails frequently not so much after the parlor was put in in 1972 but prior to that for sure.

There was something about seeing 50 Guernseys in that old stanchion barn with all their named written in chalk with the breeding and fresh dates. We had a computer called Uncle Jerry. He kept those dates in his head and kept things up to date. He also drove the hay truck in the summer while we picked up bales. In a summer we could make 15,000 bales of hay when we filled all the barns. Grass hay and pea hay were the main staples. The records made at Western Glow were based on grass hay, grass haylage and a limited amount of alfalfa hay and some purchased corn silage. The pea vine hay and some oat hay were fed to the heifers and dry cows. The pasture consisted of grass and clovers with some cover crop. Grazing is not the best way to get high milk but those 6 months of intense physical therapy add years to the life of a cow.

Housing cows in loose housing required a steady supply of shavings to keep the barns clean and dry. With the energy crisis starting in 1973 some of the mills started looking at burning the shavings to create energy to run the sawmills. It was getting more expensive to operate in the current mode. The dairy business has always been cyclical and times and technology changes. The early 1970's were those times. Grain went from $75.00 a ton to $125.00 a ton, hay went from $36.00 a ton to $72.00 a ton and milk was low also. Those years were tough and the rising price of real estate caused people to look at the options. The farm was worth 40 times what they paid for it originally. The golden years approached and the future was uncertain…

Chapter Ten
The Future in Black and White
The changing of the guard.....

As I sit here writing this with the aid of hindsight I see a tragedy of circumstance that led to the decision to sell the Western Glow Guernsey herd in the late 1970's. Jake was retiring. Ben Jr. bought his interest in the cows and starting in 1972 a select group of Holsteins had been slowly added to the farm. They were great cows that gave huge amounts of milk easily. Had we had the modified component pricing that Max Dowdy and the Guernsey Cattle Club had been talking about for years it might have made a difference.....but i doubt it. None of us truly appreciated the work that Grandpa had done for the last 50 years breeding the herd that became Western Glow. Not until I read Horace Backus's book on the Montvic herd years later, did it start to sink in. If you want cows that transmit you have to select bloodlines that transmit the kind of cow and the traits you are working towards. It is possible to breed bloodlines... it is impossible to hit the numbers on a piece of paper when the numbers change every four months when the proofs come out. Going back and seeing the foundation to this herd it is understandable which bulls worked and which ones did not. One thing I discovered long ago is that a 6 generation pedigree will not lie to you. It states the facts in black and white. You can breed a good individual with an odd ball bull but you will not get something that will transmit and breed through the static...that takes line breeding. There are more tools today than we have ever had at our disposal and it takes thought and perseverance and patience to see it through.

The bloodlines established at Western Glow made possible bulls like Western Glow Darimost, Henslee Farms V Fame, Gayoso View Top Command, Henslee Farms Viscount, Housleys Dari Fayvor, Clovelly Top Hornet, Western Glow TH Alstar and many others. By my count there were 34 All Americans in milking form from these bloodlines.

In 1977 the Guernseys at Western Glow were sold, they were sold in groups, several of which went to British Columbia, Canada. George White bought several head. There was no dispersal for the Guernseys they simply went on to new homes. At the time is seemed like progress. Looking back I see it was an opportunity that was lost to cultivate the bloodlines and continue Grandpa's life work.

The Holstein herd that was assembled over the early 1970's and the partnership cattle that were housed there created quite a stir in the Holstein arena. When the Guernseys were sold they purchased a group of heifers from the Harpain herd in California to get herd numbers up. Ben Jr. had an opportunity in Texas at his wife Layne's parents to take over their pecan ranch so the decision in the spring of 1978 was to hold a dispersal for the Holstein herd at Western Glow.

The end of an era...

In May of 1978 the Holsteins were dispersed at Western Glow Farms in Bow, Washington. The sale average was the highest in history for a dispersal in the Northwest. With all animals averaging $1925.00 per head from calves on through to cows. Cattle went to Washington, Oregon, Idaho, Montana, California, and British Columbia, Canada. My step father David Dujardin and I bought 19 head for our Baw Faw Holstein herd in Curtis Washington.

The farm was sold and the cattle were gone. It is now a potato farm that raises fresh market potatoes for the West Coast market. The barns are deteriorating rapidly the last few years as roofs fail to protect the old wood structures.

Almost 40 years ago but the memories remain.

Western Glow Farms Bow, Washington circa 1936

Western Glow Royal Laughter (7-7) 11656—617—Jr2—305C—2x
  S.: Silver Forest May Royal
  D.: Wesclair Foremost Linneta
  B.: Western Glow Farms, Inc., Bow, Wash.
  O.: Western Glow Farms, Inc., Bow, Wash.

FOR FEBRUARY 1, 1950

Chuckluck Lady Lovely (10-10) 13974—661—Sr3—305C—2x
  S.: Chuckanut Theron
  D.: Bear Canyon's Lady
  B.: L. W. Power, Burlington, Wash.
  O.: Western Glow Farms, Inc., Bow, Wash.

Present day Wandamere Farm Boring, Oregon the farm where Maryann went in 1946

§Western Glow Farms, Incorporated, Bow, Wash.			ARHT				
*†Avondale Gem	Jr3	127	Sept	1423	60	7027	321
†Forest Green of Aldarra	Jr2	158	Sept	889	42	5457	269
†Golden Light of Aldarra	Sr2	287	Aug	926	41	9577	418
*†Loop's Pride Lass	Sr3	181	Aug	876	53	7029	380
†Potlatch Faithful's Crystal	Jr4	105	Sept	1124	49	4900	217
†Potlatch Polly	Sr4	179	Sept	802	42	6837	330
*†Prairie Bloom Glenna	5	48	June	1590	78	2500	126
s. Wandamere Royalist D. Prairie Bloom Donna Bell							
†Theo-Vale Noranda	7	156	Sept	1270	60	7720	343
*†Wandamere Anson's Emma	5	202	Aug	1204	59	9362	447
*†Western Glow Bonniebee	Jr4	107	Sept	1439	62	5779	280
†Western Glow Country Flower	Jr2	75	Sept	1085	48	2892	132
s. Western Glow Maryann's Royal D. Wesclair Pansy's Flower							
†Western Glow Country Lady	Sr2	145	Sept	861	40	5394	264
†Western Glow Country Minuet	Sr2	196	Sept	757	39	6344	331
*†Western Glow Dutiful	Sr4	113	Sept	1498	65	6033	261
†Western Glow Masher's Ann	Jr2	170	Sept	879	51	5162	282
*†Western Glow Mistress	7	219	Sept	1158	59	10625	560
†Western Glow Royal Duty	Jr2	242	Sept	967	45	8910	400
†Western Glow Royal Jane	Sr2	111	Sept	994	52	4081	221
*†Western Glow Royal Maid	Sr4	102	Sept	1264	68	5097	264
*†Western Glow Royal Melba	Jr4	168	Sept	1023	42	7635	343

The Western Glow prefix (Holsteins)

Four years later after my Grandpa Ben passed away in 1982, we were privileged to attain the Western Glow prefix for our Holstein herd. We were in the early stages of acquiring some quality cattle from Bill and John Smalley and other select purchases. That herd was in existence in Curtis, Washington from 1977 to 1993. Then the herd was sold at our dispersal in July of 1993. We had had some success and sold two cows for over $11,400.00 in the mid 1980's We also bought a select group of heifers in the 1987 Hilltop Hanover Sale in Yorktown Heights, NY that turned out well. We also owned at one time **Hilltop Hanover B Della** a Sexation daughter out of **Brigeen Hanover Debra** the first 4th generation Excellent and over 40,000 milk cow in the breed. She went on to be the 5th generation Excellent and over 38,000 milk with 334,000 lifetime after she went to Ocean View Holsteins in our dispersal.

Today my children and grandchildren are using the Western Glow prefix for their Holsteins.

My son Ben had the Grand Champion Holstein and Reserve Supreme Champion at the 2015 Oregon State Fair with his 4 yr old VG 88 Atwood daughter .

l to r: Kristi Tracy, Karen Sloan, Brea Tracy, Ben Sloan, Corby Groen, Carol Young, Jodi Capini

Chapter Eleven
Western Glow - Part 2
We bought a Guernsey

With all this information discovered it the last couple of years the wheels started turning, I had a renewed interest in not only family history but also the Guernsey breed. I started paying attention to everything Guernsey and was pleasantly surprised by the buzz I was hearing from the talk about A2A2 milk to the desire of people to rediscover this great breed. If you have ever had cold Guernsey milk right out of the tank you know that nothing on this planet tastes better. I started paying attention to the sale catalogs and the Journal. What I saw impressed me and I thought there was a place for some old school genetics that emphasized 5.0% test, fertility and strength.

So we started looking for a Guernsey that would fit the bill knowing that there were not a lot of cattle to choose from particularly out West. We toured Abiqua Acres the great herd of Alan and Barbara Mann in Silverton, Oregon.

We went to the Lewis County Holstein Sale in Chehalis that weekend. There were two Guernsey heifers in the sale and we picked out the one we thought was best. The dam was VG and had 5.0% test with 934 of fat. The calf was sired by a Yogi Bear son and she looked the part. The heifer topped the sale at $3,500.00....the whole sale, Holsteins and all.

South Bank M Susie

l to r: Danner Tracy, Kirt Sloan, Luana Sloan, Brea Tracy, Earl Kegley

Susie has had quite a year and has been undefeated in the show ring in 2015.

Brea Tracy on the halter.

Junior Champion at the Oregon State Fair

Danner Tracy taking care of the barn crew

Brea Tracy on the halter

l to r: Earl Kegley, Hans Leuthold, Ben Friedrichs

Taken around 1968 at the Evergreen Classic Sale in Mt Vernon, Washington. Who knew 47 years later that we would buy our first Guernsey from Earl Kegley in Chehalis, Washington.

South Bank M Susie

Going Old School....

This spring after we bought Susie we attained the Western Glow prefix for the Guernseys and in keeping with my old school roots started looking for some semen to breed Susie to this coming spring. I put on Facebook that I was looking for some **Western Glow Darimost** semen and with in a couple of days I found some...22 units of Darimost, 5 units of **Housleys Dari Fayvor,** 1 unit of **Clovelly Top Hornet** so we are set for a while....going to need some more Guernseys first.

Susie is housed at Ben's place in Curtis, Washington.

This has been a labor of love to finally get the words and pictures down on paper, even though it is digital. I hope you can see and understand how methodically the Western Glow breeding program was put together, and how those philosophies still are applicable today even in the age of genomics.

There is a stunning similarity to the great breeder herds. Montvic, Skagvale, Colony Farms, Carnation, Roybrook, McDonald Farms, Langwater Guernseys all line bred with a program to improve their herds and the breed. Line breeding plays a huge role in those that succeed in the long run. The successful breeders look ahead to see what the breed needs ten, fifteen, twenty years down the road and has the goods ready to go at that intersection. Choose your goals, stay in the rut, and the winding road will come back in line with your goals as the pendulum swings back and forth.

One thing to remember about breeding cows....the last genetic impact you have on an dairy animal is when you deposit the semen in that animal's dam. The blood that runs through their veins never changes, no matter what is on the paperwork.

Thank You

There are many people that I have been fortunate to meet in my years in this business. Some have taken the time to talk and help me figure out a little more about breeding cows and doing things the right way.

Grandpa Ben -It has been 33 years since you have been gone and there is not a day that goes by that I don't miss you. You taught me the value of a good cow and how to make it happen with the cows you have. Every mating is an opportunity to make something better. And there is no greater satisfaction than finding a great cow before she is great.

Ray Nelson - taught me to sell a cow when she was at the peak of her performance. Those are the kind the market wants and will pay a premium for. Sell the hot ones, keep the good ones.

David Younger - He was like talking to Babe Ruth. Best marketing guy I ever met and absolute perfection in the way the cows were cared for at Hilltop Hanover. He was a great mentor and a great friend. I asked him how much corn he fed, his reply "Oats are for cows, Corn is for Hogs" Totally true.

Doug Maddox - His gift was this "put your money in good cows, I have never seen a tractor have a calf"

Marvin Nunes - The master at putting cow families together and getting lifetime production. We got to work with some of those families. The Ocean View catalogs were amazing to study.

Pete Blodgett - Best commercial cow man ever! He loved cows that could eat and make milk. Taught me about health traits before they were cool. Taught me the way to measure strength was looking from behind the cow look at the width of the lower bone in the front leg, substance of bone matters.

Pete DeHaan - My best friend and mentor, takes doing things right to another level. Never have I seen as much passion about this business as he brings to the table.

Horace Backus - Taught me to love history and how herds get put together in his books about Dunloggin and Montvic....but mostly Montvic. The story of Montvic is the reason A.I. and Holsteins ruled the day for the past 80 years.

Cliff Shearer - A friend on Facebook that shared his love of history and breeding and the data archives at CDN that allowed me to put the pieces of the puzzle together. And he told me about the Langwater book!

Brent Clements - One of the key modern day Guernsey enthusiasts that promotes the breed with his heart and soul. The future is looking great for Guernseys.

Kevin Gass - The guy that owns a hell of a Guernsey cow in Australia that I am deeply enamored with. Koala is a beast!!

David Dujardin - My step Dad and partner for many years, who took a chance on an 18 year old kid and bought a dairy farm in Chehalis, Washington and put me in charge. Years later I asked him…"What the hell were you thinking?" I learned a lot there. You can start out with below average cows ("and we did, they were cheap and terrible") but you have to find some foundation females that work if you want to make progress. Seeing a herd develop and seeing the effects both good and bad in selecting genetics and discovering cow families was a life's lesson I will never forget.

Dedication

"Dairy farming is an incurable mental illness"

And breeding purebred cattle is worse. I have been in remission a time or two but never been cured. As the saying goes 'Being addicted to cows makes sure you don't have enough money to be addicted to drugs.' I say this jokingly but those in the business know the saying is somewhat true.

One of the best things that ever happened to me was when my wife Luana Sloan knocked on my front door in July of 2008. The first time we met we talked for 4 hours and we have never quit talking. She has supported me in every way and been the best friend I have ever had. Your support through all of this endeavor means everything to me. As the sign in our living room says "I wish I would have met you sooner so that I could love you longer."

To my kids Karen, Kristi, and Ben. You will never forgive me for selling the cows in 1993 and I totally understand and appreciate those feeling because I have them too. Your love and support over the years even though we have been separated by the miles means everything to me. I love you all!! And thanks for taking care of "the guerndog" and letting us show with you!

To Brea and Danner thank you for taking care of Susie and making her look good on show day. You guys are gifted showmen and the most perfect Grandkids!!

To all the rest of the kids and grandkids in Idaho we love you and thank God we all get along!! Love you a lot!!

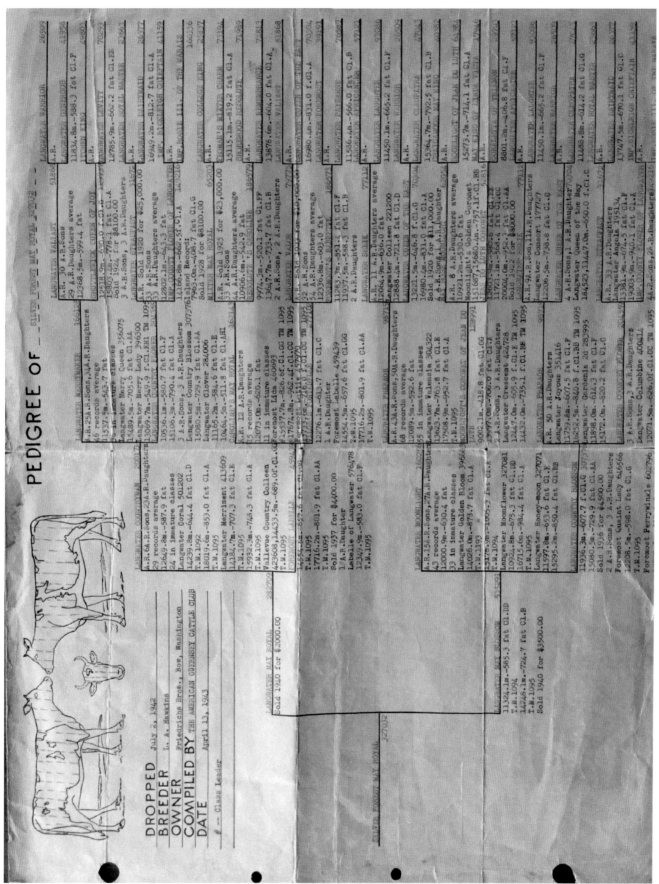

The pedigree of Silver Forest May Royal

The pedigree of McDonald Farms Developer

The pedigree of Fritzlyn Jeanette's Flash

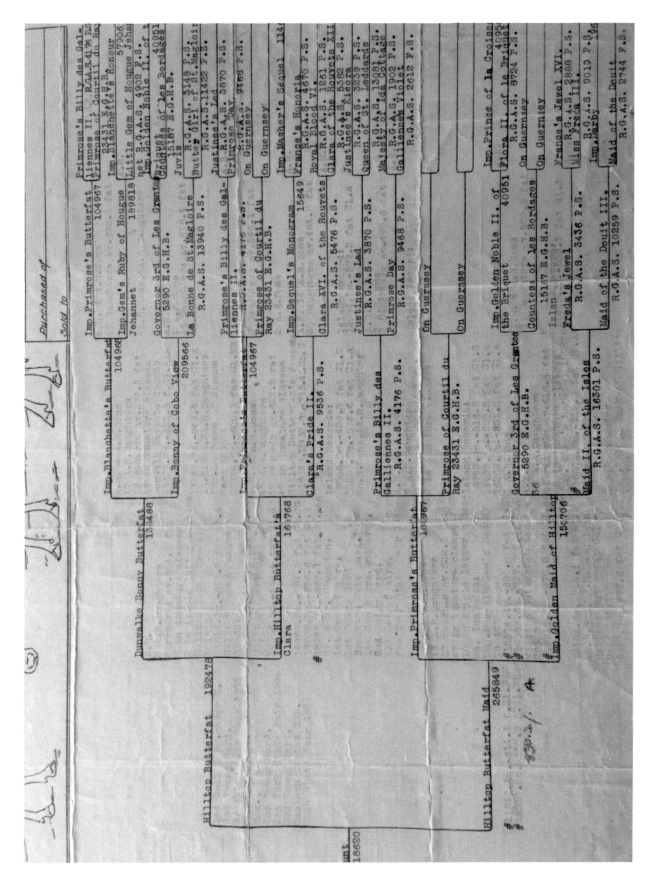

The pedigree of Hilltop Butterfat Count

Hilltop Farm in Connecticut

WESTERN GLOW FOND GAY DUCHESS

AMERICAN GUERNSEY ASSOCIATION

```
WESTERN GLOW FOND GAY DUCHESS    USA000001920224                                              Current Date: 1-7-2015
      Born: 01-05-1959                              WESTERN GLOW M B DUCHESS           87 @ 9-02
   Breeder: WESTERN GLOW FARMS INC                  7-04 308D 2X  13630M 6.0%   814F                         DHIA
           BOW                WA 98232              WESTERN GLOW LADY MARY DUCHESS     84 @11-00
     Owner: ANDERSON EDWARD C EST                   8-00 306D 2X  15410M 5.5%   844F                         DHIA
           BUSHY PARK FARM
           P O BOX 98                                     AGED                  1964 AA
           WAKE               VA 23176                    4 YRS                 1963 AA
 Appraisal: 94 @ 8-10                                     AGED                  1965 RES AA
  2-02  305D 2X  13296M 4.9%   645F        DHIR           S & G CH              1964 N G S
        309D 2X  13432M 4.9%   653F        DHIA      1st  AGED                  1964 N G S
  3-02  365D 2X  15091M 5.5%   825F        DHIA      2nd  AGED                  1965 N G S
  4-05  305D 2X  17305M*5.5%   958F*       DHIR      1st  4 YRS                 1963 N G S
  5-05  305D 2X  10937M 6.0%   656F        DHIR           BEST UDD CH           1964 N G S
        314D 2X  11118M 6.0%   668F        DHIA      2nd  BEST UDD 4 YRS        1963 N G S
  6-05  305D 2X  14345M 5.1%   734F        DHIR      1st  BEST UDD AGED         1964 N G S
        309D 2X  14477M 5.1%   742F        DHIA           S & G CH              1963 N J
  7-07  305D 2X  17180M 5.3%   916F        DHIR           S & G CH              1964 E GUERNSEY
        360D 2X  18890M 5.3%  1008F        DHIA           S & G CH              1965 N J
  8-08  305D 2X  16900M 5.1%   858F        DHIR      1st  AGED                  1964 E GUERNSEY
       2662D    116869M 5.4%  6276F 0.0%  0P  LIFE   1st  AGED                  1965 N J
                                                     1st  4 YRS                 1963 N J
                                                     2nd  PRODUCE               1965 W WASH

                                                    FRITZLYN FLASH              USA000000411629
                                                    ----------------------------------------------------
FRITZLYN JEANETTES FLASH          USA000000475548  -FRITZLYN GYPSY KINGS JEANETTE   USA000000969573
-------------------------------------------------------
12/14 USDA PTA  -2316M +0.21%  -74F +0.02%  -73P
  +4.6DPR -1.2PL  -407NM$   0NM%ile -400CM$   95% Rel
     AGA 12/14  PTAT -3.4 -1.1UDC -2.0FLC 63%Rel  PTI -298

   ST   SR   BD   DF   RA   TW   RV   RL   FA
  -6.1 -4.2 -3.8 -3.7 L1.7 -3.7      0.0 L1.4
   FU   RH   RW   UC   UD   TP   TL
  -0.9 -2.4 -2.6 -1.2 S0.1 W0.8 -1.2               WESTERN GLOW BUTTERFAT TRAVLER  USA000000511205
                                                   ----------------------------------------------------
WESTERN GLOW TRAVLERS DUCHESS     USA000001682143  12/14 USDA PTA  -2616M +0.13% -100F +0.10%  -71P
-------------------------------------------------    +7.5DPR -2.6PL  -546NM$   0NM%ile -524CM$   79% Rel
Appraisal: 86 @ 12-11
  2-02  305D 2X  10081M 5.3%   535F        DHIR
  3-02  303D 2X  10982M 5.7%   628F        DHIR

  4-02  305D 2X   9911M 5.2%   520F        DHIR
  5-05  326D 2X  11563M 5.2%   602F        DHIA
  6-06  311D 2X  10588M 5.3%   560F        DHIA
  7-06  300D 2X   9780M 5.0%   489F        DHIR
  8-09  317D 2X  11980M 6.1%   727F        DHIA
 10-01  293D 2X  12010M 5.7%   690F        DHIR
 11-03  279D 2X  11120M 5.3%   594F        DHIR
WESTERN GLOW TRAVELERS DUANN       91 @ 5-09
  5-00  305D 2X  13370M 5.2%   692F        DHIR
WESTERN GLOW NOBLE GAY DUCHESS     90 @ 7-01
  5-05  310D 2X  16010M 5.3%   854F        DHIA
WESTERN GLOW NOBLE DUTCH GIRL      83 @ 7-04
  5-02  315D 2X  16350M 5.5%   905F        DHIA
```

The pedigree of Western Glow Fond Gay Duchess

The cover:

l to r: Kirt Sloan, Ben Friedrichs, Kent Sloan

This picture was taken in the milking barn at Western Glow Farms Bow, Washington in August of 1963.

This was the day Western Glow Fond Gay Duchess was loaded to head to Woodacres in Princeton, New Jersey. She was trucked to Burlington, Washington and loaded on a box car for the train trip east. Grandpa and John Rietman from Lynden, Washington made the trip with her to New Jersey then flew home to Seattle.

She was:

All American 4 yr old 1963

All American Aged Cow 1964

Res. All American Aged Cow 196

The Author:

Kirt Sloan was born in Mt. Vernon, Washington and grew up in Skagit County spending as much time as possible on the farm. Cows, reading and history were always key interests. He loved hanging out with adults as the stories and jokes were always better with grown ups. He graduated from Burlington Edison High School in 1976 and attended Cal Poly State University in San Luis Obispo, CA for a short time before taking over the management of Wildwood Dairy in Curtis, Washington in partnership with David Dujardin at the ripe old age of 18 in 1977. Over the next 16 years he developed a herd of registered Holsteins purchasing select animals from Montana, and the Pacific Northwest. A key group of heifers were purchased from Hilltop Hanover Farm in Yorktown Heights, NY in 1987. The herd exported animals to Saudi Arabia, Taiwan, and Japan. The herd was dispersed in 1993.

Kirt has worked in the dairy equipment business for the last 22 years and took a two year break in 2004 to work as a herd manager for Pete DeHaan Holsteins in Salem, Oregon. He currently is the Robotic Sales Specialist for the West Region for DeLaval, Inc.

He lives with his wife Luana in Twin Falls, Idaho and together they have 7 children and 8 grandchildren. They enjoy traveling and camping with their 2 golden retrievers. They currently own 1 Guernsey heifer.

email: farmerkirt@icloud.com

Phone: 208-404-9046

Made in the USA
Las Vegas, NV
21 September 2021